T0254985

Lecture Notes in Computer Science 13745

Founding Editors

Gerhard Goos
Karlsruhe Institute of Technology, Karlsruhe, Germany

Juris Hartmanis
Cornell University, Ithaca, NY, USA

Editorial Board Members

Elisa Bertino
Purdue University, West Lafayette, IN, USA

Wen Gao
Peking University, Beijing, China

Bernhard Steffen
TU Dortmund University, Dortmund, Germany

Moti Yung
Columbia University, New York, NY, USA

More information about this series at https://link.springer.com/bookseries/558

Wenjuan Li · Steven Furnell · Weizhi Meng (Eds.)

Attacks and Defenses for the Internet-of-Things

5th International Workshop, ADIoT 2022 Copenhagen, Denmark, September 30, 2022 Revised Selected Papers

Editors
Wenjuan Li
Hong Kong Polytechnic University
Hong Kong, China

Steven Furnell
University of Nottingham
Nottingham, UK

Weizhi Meng
Technical University of Denmark
Kongens Lyngby, Denmark

ISSN 0302-9743 ISSN 1611-3349 (electronic)
Lecture Notes in Computer Science
ISBN 978-3-031-21310-6 ISBN 978-3-031-21311-3 (eBook)
https://doi.org/10.1007/978-3-031-21311-3

© The Editor(s) (if applicable) and The Author(s), under exclusive license
to Springer Nature Switzerland AG 2022
This work is subject to copyright. All rights are reserved by the Publisher, whether the whole or part of the material is concerned, specifically the rights of translation, reprinting, reuse of illustrations, recitation, broadcasting, reproduction on microfilms or in any other physical way, and transmission or information storage and retrieval, electronic adaptation, computer software, or by similar or dissimilar methodology now known or hereafter developed.
The use of general descriptive names, registered names, trademarks, service marks, etc. in this publication does not imply, even in the absence of a specific statement, that such names are exempt from the relevant protective laws and regulations and therefore free for general use.
The publisher, the authors, and the editors are safe to assume that the advice and information in this book are believed to be true and accurate at the date of publication. Neither the publisher nor the authors or the editors give a warranty, expressed or implied, with respect to the material contained herein or for any errors or omissions that may have been made. The publisher remains neutral with regard to jurisdictional claims in published maps and institutional affiliations.

This Springer imprint is published by the registered company Springer Nature Switzerland AG
The registered company address is: Gewerbestrasse 11, 6330 Cham, Switzerland

Preface

The 5th International Workshop on Attacks and Defenses for Internet-of-Things (ADIoT 2022) was held on September 30 with ESORICS 2022 in Copenhagen, Denmark. Due to the COVID-19 situation, it was held in hybrid mode.

The Internet of Things (IoT) technology has been widely adopted by the vast majority of businesses and is impacting every aspect of the world. However, the natures of the Internet, communication, embedded OS, and backend recourses make IoT objects vulnerable to cyber attacks. In addition, most standard security solutions designed for enterprise systems are not applicable to IoT devices. As a result, we are facing a big IoT security and protection challenge, and there is an urgent need to analyze IoT-specific cyber attacks to design novel and efficient security mechanisms. This workshop focuses on IoT attacks and defenses, seeking original submissions that discuss either practical or theoretical solutions to identify IoT vulnerabilities and IoT security mechanisms.

This year, ADIoT received 18 submissions, and each submission was reviewed by at least three reviewers in a single blind process. According to the novelty and quality, seven regular papers were accepted along with three short papers, resulting in an acceptance rate of 38.9%. For the conference program, we had three paper sessions and one keynote: Zhiqiang Lin (The Ohio State University, USA) addressed "Rethinking the Security and Privacy of Bluetooth Low Energy".

For the success of ADIoT 2022, we would like to first thank the authors of all submissions and all the Program Committee members for their great efforts in selecting the papers. We also thank all the external reviewers for assisting the reviewing process.

September 2022

Wenjuan Li
Steven Furnell
Weizhi Meng

Preface

Organization

General Chair

Anthony T. S. Ho	University of Surrey, UK

Program Co-chairs

Wenjuan Li	Hong Kong Polytechnic University, China
Steven Furnell	University of Nottingham, UK
Weizhi Meng	Technical University of Denmark, Denmark

Technical Program Committee

Mohiuddin Ahmed	Edith Cowan University, Australia
Claudio Ardagna	Università degli Studi di Milano, Italy
Ali Ismail Awad	Lulea University of Technology, Sweden
Silvio Barra	University of Naples Federico II, Italy
Chao Chen	Swinburne University of Technology, Australia
Elena Doynikova	SPC RAS, Russia
Reza Malekian	Malmo University, Sweden
Jianbing Ni	Queen's University, Canada
Evgenia Novikova	Saint Petersburg Electrotechnical University, Russia
Luca Ferretti	University of Modena and Reggio Emilia, Italy
Jianming Fu	Wuhan University, China
Yunguo Guan	University of New Brunswick, Canada
Georgios Kambourakis	University of the Aegean, Greece
Kwok Yan Lam	Nanyang Technological University, Singapore
Jay Ligatti	University of South Florida, USA
Vincenzo Moscato	University of Naples, Italy
Claudia Peersman	University of Bristol, UK
Josef Pieprzyk	CSIRO/Data61, Australia
Indrajit Ray	Colorado State University, USA
Jun Shao	Zhejiang Gongshang University, China
Gang Wang	University of Connecticut, USA
Xuyun Zhang	Macquarie University, Australia
Cong Zuo	Nanyang Technological University, Singapore

Steering Committee

Steven Furnell	University of Nottingham, UK
Anthony T. S. Ho	University of Surrey, UK
Sokratis Katsikas	Norwegian University of Science and Technology, Norway
Weizhi Meng (Chair)	Technical University of Denmark, Denmark
Shouhuai Xu	University of Texas at San Antonio, USA

Subreviewers

Qingqing Gan
Florian Gondesen
Wenzhuo Yang
Xu Yang
Fei Zhu

Contents

The Final Round: Benchmarking NIST LWC Ciphers on Microcontrollers

Sebastian Renner[1,2](✉), Enrico Pozzobon[1,3], and Jürgen Mottok[1]

[1] OTH Regensburg, Regensburg, Germany
{sebastian1.renner,enrico.pozzobon,juergen.mottok}@othr.de
[2] Technical University of Munich, Munich, Germany
[3] University of West Bohemia, Pilsen, Czech Republic

Abstract. In this work, we present our benchmarking results for the ten finalist ciphers of the Lightweight Cryptography (LWC) project initiated by National Institute of Standards and Technology (NIST). We evaluate the speed and code size of various software implementations on five different platforms featuring four different architectures. Moreover, we benchmark the dynamic memory utilization of the remaining NIST LWC algorithms on one 32-bit ARM controller. We describe our test cases and methodology and provide some information regarding the design and properties of the finalists before showing and discussing our results. Altogether, we evaluated almost 300 implementations of the 3rd round candidates and pick the most appropriate and best (primary) implementation of each cipher for our comparisons. We include a variant of AES-GCM in our benchmarking in order to be able to compare the state-of-the-art to the novel LWC ciphers. Our research gives an overview over the performance of the latest software implementations of the NIST LWC finalists and shows under which circumstances which candidate is performing the best in our individual test cases. Additionally, we make all benchmarking results, the code for our test framework and every tested implementation available to the public to ensure a transparent testing process.

Keywords: Lightweight cryptography · Benchmarking · Embedded systems · Evaluation framework

1 Introduction

The progress in the IoT branch in the recent years led to a significant growth of internet-connected and "smart" devices. With more embedded systems communicating over the internet in both the private and professional domain, new security and privacy challenges arise. Especially devices which process privacy related data or operate in critical environments require appropriate protection against eavesdropping and other attacks. Parts of these requirements are usually fulfilled by utilizing well-established cryptographic algorithms, e.g. the Advanced Encryption Standard (AES). However, in some cases specifically efficient cryptographic primitives might be needed due to the limited computing resources

© The Author(s), under exclusive license to Springer Nature Switzerland AG 2022
W. Li et al. (Eds.): ADIoT 2022, LNCS 13745, pp. 1–20, 2022.
https://doi.org/10.1007/978-3-031-21311-3_1

of small embedded systems. This typically concerns sensor networks, wearables, smart meters or RFID chips. Depending on the specific use case, such systems require either very high-performing, low-memory-consuming or energy-efficient firmware that still delivers a certain security level. The challenge of enabling constrained devices to securely transmit and receive data has been addressed by academia and in the industry in the last couple of years. Moreover, in 2018 NIST issued a call for algorithms that support authenticated encryption with associated data (AEAD) modes and published a report regarding "Submission Requirements and Evaluation Criteria for the Lightweight Cryptography Standardization Process" [32]. This marked the start of a LWC standardization process directed by NIST, to which initially 56 algorithms have been accepted as 1st round candidates. After various evaluation rounds and (external and internal) research, NIST announced ten finalist algorithms in March 2021. Currently the NIST LWC project is in its last selection phase and the announcement of a "winner" primitive is expected before the end of 2022. This algorithm will later be standardized in a Federal Information Processing Standard (FIPS) for the use in specific low-power environments.

In this work, we shortly present our benchmarking framework, custom-made for evaluating software implementations of LWC ciphers on different microcontroller unit (MCU) platforms. We then show our latest results when comparing the NIST finalists in our test setup. More specifically, we present measurements for the speed, the Read Only Memory (ROM) and the Random Access Memory (RAM) utilization of in total 295 implementations on (up to) five different MCUs. We provide benchmarking figures for the most recent and optimized implementations of the NIST LWC finalists.

Contribution. First, we built a tailored Hardware in the Loop (HIL) benchmarking framework for evaluating the performance of software implementations of the NIST LWC candidates. We explain our testing methodology, platform integration and standardized test cases from which we obtain detailed measurements for speed, size and RAM utilization in an automated manner. Second, we host all the source code of the tested implementations and our own framework on our publicly available website[1,2]. Moreover, every test result is made available through our web page upon completion of the test process. For each test, we store and show not only the benchmarking results but also metadata (time/date of last implementation change, time/date of test execution etc.) to guarantee maximum transparency[3]. Furthermore, we provide a public submission server to which designers can submit their latest implementations which then get tested automatically and are integrated into our result database. Lastly, we present a subset of our testing results for the ten finalist algorithms of the NIST LWC project. In total, we evaluated the speed, ROM and RAM consumption of 295 different implementations on up to five different MCUs. Besides comparing the

[1] https://lab.las3.de/gitlab/lwc/compare.
[2] https://lab.las3.de/gitlab/lwc/candidates.
[3] https://lwc.las3.de.

implementations to each other, we confront the LWC ciphers to the current standard for AEAD – AES-GCM – and show which implementations deliver the best performance in our tests. Besides raw performance figures, we also provide some information on "soft" factors that might influence the decision making when selecting LWC ciphers. For example, we disclose how many different implementations we received and were able to benchmark for each NIST LWC finalist.

Related Work. There exists various research on benchmarking cryptographic algorithms, both within and outside of the NIST LWC project. Dinu et al. built the FELICS framework to support the fair evaluation of software implementations of block ciphers. They benchmarked the performance of 19 block ciphers in their work [19,20]. Later Cardoso et al. extended the tool to support AEAD modes [30]. Weatherley contributed multiple software implementations of the NIST LWC ciphers which have been specifically optimized for AVR architecture. Moreover, he conducted speed benchmarks on 32-bit and 8-bit Arduino platforms [36]. Further benchmarking experiments have been carried out by members of the NIST LWC team themselves. They tested the performance of 2nd round candidates and measured the code size of lots of implementations on various different MCUs [28]. Campos et al. specialized on evaluating the LWC ciphers on a single architecture – RISC-V. In their work, the authors focus on comparing the performance of implementations written in plain C to assembly-optimized versions for RISC-V [16]. Cazorla et al. also analyzed the performance of cryptography on a dedicated platform. They evaluated 17 block ciphers on a TI MSP430 MCU in 2015 [17]. A popular benchmarking framework in this domain is SUPERCOP, which has been developed and used already prior to the NIST LWC project [9]. SUPERCOP supports a wide variety of testing platforms, including desktop and (higher end) embedded systems. Ankele et al. also conducted performance tests on desktop CPUs. They evaluated 2nd round candidates of the CEASAR competition, which was a different competition for authenticated encryption algorithms held prior to the efforts from NIST [1,15]. Additional examples for benchmarking software implementations of ciphers include the works of Tschofenig et al., who studied the performance of elliptic curve cryptography on ARM Cortex-M ciphers and Hyncica et al., who compared multiple lightweight metrics of 15 symmetric ciphers on different embedded platforms [26,35].

Outline. The rest of this paper is structured as follows: Sect. 2 provides information on how our framework is built, how we designed our test cases and how we measure the different performance metrics. Additionally, this section describes the various MCUs that form our devices under test. Section 3 introduces the ten remaining LWC ciphers from the final selection round and shows the properties of each candidate. We also touch on the different modes of the ciphers and share some information regarding the amount of available and analyzed software implementations. In Sect. 4, we present the results of our evaluations. We compare the candidates to each other and show which implementations perform best in which category (in our tests). This section also discusses our results. Section 5 concludes the paper with some closing remarks.

2 Framework

When we started planning our benchmarking project, we first evaluated the requirements for the testing framework. We investigated both strategies, using or extending an existing tool or building a new one. Three requirements have been crucial for us: First, we wanted to be able to execute the performance measurements on real hardware that can be replaced and extended if needed. Second, the testing procedure should be as standardized and automated as possible. This is especially important to a) generate reproducible results b) be able to test many ciphers fast c) ensure tests can be conducted and results can be updated throughout the whole duration of the NIST LWC project. Third, we wanted to aim for maximum transparency by sharing the source code of our framework, keeping a repository of all tested implementations and publishing updated results once they are available.

After evaluating software at the time of building the toolchain, we found that extending an existing framework would still require lots of work and we then could still not tailor the code enough to perfectly fit our requirements. That is why we made the decision to implement our own benchmarking tool, custom-made for the NIST LWC project. Right now we operate five different MCUs that represent four different architectures. For each of the test platforms we provide a dedicated template that mostly initializes the device and acts as a runtime environment for the tests. The test routines are started and monitored by generic scripts that are the same for each MCU. This structure allows us to easily integrate new hardware platforms with minimal effort. We expect the to-be-tested implementations to define certain functions and with an API that was specified by NIST. The compilation, flashing, testing and results collecting procedures are handled by the framework in an automated manner, once an implementation is processed as a new input to the toolchain. After ensuring that the tests have been executed without an error, the results are made publicly available on our website. For more details regarding the used tools and the exact compilation and test process, see Sect. 3 of [29].

2.1 Tested Metrics

We evaluate each implementation on the basis of three metrics: The time an algorithm takes to process (encrypt/decrypt/authenticate/verify) a predefined set of data, the size of the binary of the implementation and the consumption of RAM on the device under test. These performance indicators have been used by related work in the past and can be seen as the most important metrics when it comes to the performance evaluation of lightweight software implementations [20]. However, not every metric might be equally important for every use case. In environments with high real-time constraints, speed can be the highest priority, while in other low-cost scenarios a minimal code size could be more important. Moreover, highly optimizing an implementation for all three metrics is not possible since e.g. usually high speed implementations come with a greater code size. Evaluating all of these metrics allows us to study which effects optimizations of a specific metric can have on the others, so one can choose an appropriate implementation according to the constraints of the use case.

Execution Time. NIST required the designers to submit a text file with static test vectors for their algorithms in the submission package. These test vectors files can be generated by using a C-tool provided by NIST. The test vectors consist of 1089 different input/output data pairs that allow for checking if an algorithm functions as intended. The test data includes various cases, ranging from lower to higher length plaintexts with or without associated data to be authenticated. We use these 1089 data pairs to measure the speed of an implementation. In our speed test case, every algorithm has to encrypt and decrypt all of the data available in the test vector file of the algorithm. This way we cannot only evaluate the execution time for different input data but we can also make sure that the submitted implementation works correctly. We use a logic analyzer and pin toggling on the MCUs to measure the time for each encryption/decryption. If the test vector contains associated data, the time for the authentication is also taken into account. Due to the design nature of the symmetric LWC ciphers, the encryption time generally equals the decryption time. In the end we can calculate an average processing time for the NIST test vectors which gives us a comparable measurement for any implementation. While benchmarking the performance within a dedicated IoT protocol would also be interesting for practical use cases, we believe choosing the predefined test vectors allows us to make a generic and broad evaluation of all the implementations.

Code Size. To measure the size of the binary of an implementation, we compile each algorithm such that it can be used and flash on the target devices. In the compilation process, we define the flags according to what NIST recommended it the call for submission [32]. Of course, the initialization of the MCU in its template already takes up some space in the compiled binary. To make sure that we only measure the overhead code size of the actual implementation, we first determine the size of our runtime environment on each MCU and can then calculate the difference between that and the total binary size to get the exact code size of the implementations per platform.

RAM Utilization. In order to determine the RAM utilization, the memory of the device is filled with a predefined pattern. Then, the test vectors for the implementation are processed and the RAM is inspected afterwards. We observe which memory regions remained untouched and again take the already used memory by the template into account. This delivers a comparable number of utilized bytes during the execution of the tested algorithm.

2.2 Test Platforms

We aimed for covering a wide range of architectures and chips used in the IoT domain. Initially, we did not feature a RISC-V controller but we integrated the Sipeed MCU later due to the rising interest in RISC-V. This way we could show that the modularity of the framework makes it easy to switch out or add a target device.

Arduino Uno R3. The Arduino Uno is based on the ATmega328P 8-bit AVR MCU [2]. It operates on a clock frequency of 16 MHz and features 32 KB of flash, 2 KB SRAM and 1 KB EEPROM memory. The RISC architecture offers a pipeline with two stages and 32 general purpose registers. The ATmega328P represents the low-cost/low-performance end of our supported MCUs.

STM32F1 "bluepill". The evaluation boards commonly referred to as "blackpill" or "bluepill" are lower-end platforms equipped with a STM32F103C8T6 MCU which is based on an ARM Cortex-M3 chip [33]. The 32-bit core works at a maximum clock of 72 MHz and incorporates 64 KB of flash memory. The ARMv7 architecture implements a three-stage pipeline and the Thumb(-2) instruction set with 16/32-bit instructions.

STM32 NUCLEO-F746ZG. This NUCLEO board is based on a high-power Cortex-M7 chip clocking at 216 MHz [34]. Moreover, it can store up to 1 MB of data in its flash memory, therefore, it is suited for more demanding IoT use cases. The core implements a six-stage pipeline and includes basic branch prediction. This evaluation platform is part of the upper performance bracket, together with the two following devices.

Espressif ESP32 WROOM. The Espressif board features a 32-bit Xtensa LX6 MCU with a clock frequency of 240 MHz and 4 MB of flash memory [23]. It has a pipeline with five stages and supports 82 basic RISC instructions. While the majority of optimized implementations still seems to be developed in ARM assembly, the ESP32 with its Xtensa core is very popular in the maker community and many IoT projects are built upon the LX6 core.

Sipeed Maixduino RISC-V 64. This evaluation board is a very powerful RISC-V-based controller. It contains the Kendryte K210 64-bit MCU which can be clocked up to a frequency of 400 MHz and has 8 MB of flash storage [31]. One could argue that this platform is too powerful to be a target for LWC, however, it still allows us to evaluate the algorithm implementations on a 4th architecture, RISC-V, which is an emerging instruction set that many future IoT products might make use of.

3 Evaluated Ciphers

We started our benchmarking efforts shortly after the beginning of the NIST LWC project and conducted our first tests already on the 56 first round candidates. At that time, mostly reference implementations of the variants have been available and competitive comparability was limited. With the reduction of the algorithms for the 2nd and 3rd round, the focus obviously concentrated on the ciphers that were and are still in the competition. Consequently, more (optimized) implementations became available for the remaining candidates which can then lead to a more meaningful performance comparison. Optimally, we would have implementations highly optimized towards different metrics for each cipher and different architectures. We will show in the following that is not the case for every candidate due to various reasons. In the upcoming paragraphs, we will shortly introduce the key properties of each finalist. This should provide an overview over available modes, variants and design principles of the ciphers. Note that we do not discuss the hash modes of any LWC candidate (if present) since we only consider the AEAD test case in our benchmarking. Besides providing some details regarding the ciphers' designs, we also mention how many implementations we received and tested for each finalist.

ASCON. The AEAD mode of ASCON mainly works on the basis of a monkey-duplex construction [12]. ASCON already took part in the CAESAR competition and was selected as part of the winning portfolio for lightweight authenticated encryption in 2019 [15]. The algorithm offers 128-bit security and uses a 320-bit permutation. There exist three variants in the ASCON suite. ASCON-128 and ASCON-128a represent the recommended types for authenticated encryption, the third variant, ASCON-80pq, offers higher resistance against quantum key-search. The number of rounds for the ciphers is expressed in a three-value vector (x, y, z) with x being the number of rounds during initialization, y being the number of rounds during message processing and z being the number of rounds at the finalization [22]. At the time of writing this paper, we have access to and have benchmarked 111 implementations of ASCON.

Elephant. Elephant is a permutation-based AEAD cipher that uses an encrypt-then-MAC construction based on a nonce. Depending on the instance, Elephant permutates a 160, 176 or 200-bit state. The 160/176-bit variants (called Dumbo and Jumbo) use the Spongent permutation, while the 200-bit version (named Delirium) uses KECCAK internally [11,14]. Dumbo achieves a security level of 112-bit, Jumbo and Delirium both reach 127-bit security [13]. We received and included 9 software implementations in our evaluation.

GIFT-COFB. GIFT-COFB is based on the block cipher GIFT-128 and the AEAD mode COmbined FeedBack (COFB) [4]. The candidate is mostly designed to perform well in a hardware setting but there also exist optimized software implementations. The submission package consists of only one variant. Our performance benchmarking includes results for six implementations of which three are specifically optimized to suit certain use cases on an ARM Cortex-M core.

Grain-128AEAD. A first version of Grain already has been a finalist in the hardware category of the eSTREAM project in 2008. The project was run from 2004 to 2008 and had the goal to promote lightweight stream ciphers for widespread use [3]. Grain-128AEAD is based on a bit-wise feedback shift register. For the 3rd round of the NIST LWC project, the Grain team tweaked the initialization of their algorithm slightly [25]. The new variant is denoted as Grain-128AEADv2 in our data. In total, we evaluated five different software implementations of Grain-128AEAD.

ISAP. ISAP is a cipher suite that provides resistance against common attacks by design. It was developed for use cases in which a system is physically accessible by an attacker and should still withstand certain attacks. These attacks include statistical (ineffective) fault attacks, differential fault attacks and differential power analysis. ISAP follows an encrypt-then-MAC scheme and uses a 320-bit ASCON or 400-bit KECCAK permutation depending on the variant. Similar to ASCON, for ISAP the number of rounds is denoted in a vector (w, x, y, z), specifying values for different phases of the algorithm [21]. Our database contains results for 33 implementations of ISAP.

PHOTON-Beetle. PHOTON-Beetle builds upon the 256-bit permutation PHOTON [24]. Similar to GIFT-COFB, it also uses a (modified) COFB mode for AEAD together with the sponge construction Beetle. PHOTON-Beetle features two different variants that differ in their absorbing/squeezing rates of the sponge [5]. For this cipher we benchmarked 14 available implementations.

Romulus. Romulus is based on the tweakable block cipher SKINNY [8]. The submission package contains three AEAD cipher groups: Romulus-N, which uses a combined feedback mode, Romulus-M, which uses a MAC-then-encrypt scheme and Romulus-T, which is based on TEDT [10]. Romulus-M offers resistance against nonce-misuse and Romulus-T is designed as a leakage-resilient mode [27]. Each cipher group specifies different variants. We have evaluated 50 software implementations of the candidate Romulus.

SPARKLE. SCHWAEMM is the AEAD instance of SPARKLE that incorporates the SPARKLE permutation. The SPARKLE permutation relies on the use of multiple Alzette boxes. Alzette is a 64-bit ARX block cipher, meaning it is consisting only of a series of additions, rotations and exclusive-or operations [6]. SCHWAEMM is based on a sponge-duplex construction with feedback. The primary variant of SCHWAEMM utilizes SPARKLE in the 384-bit version [7]. We processed 37 implementation variants of SCHWAEMM in our framework.

TinyJAMBU. TinyJAMBU represents a smaller variant of the JAMBU scheme, which was a 3rd round candidate in the CAESAR competition [37]. It specifies three variants with key sizes of 128, 192 or 256 bits. The permutation uses a nonlinear feedback register and a NAND operation. TinyJAMBU has been tweaked during the LWC project such that it provides a greater security margin against differential forgery attacks. This was achieved by increasing the number of rounds of the permutation [38]. Including both versions of TinyJAMBU, we evaluated 18 software implementations of this candidate.

Xoodyak. Xoodyak is built upon the 384-bit permutation Xoodoo, which is operated in the mode Cyclist. The design is inspired by the KECCAK permutation and also uses a sponge-duplex construction [18]. The Xoodyak team provided updated implementations of the 3rd round version of Xoodyak. Counting all versions, we received and provide results for 12 implementation variants.

4 Results

In the following we present some condensed results from our finalist benchmarking. We show plots with performance figures for the ciphers' speed (see Figs. 1, 2, 3, 4, 5) and their code size (see Figs. 6, 7, 8 and 9 and 10) on five different platforms and provide RAM utilization numbers on the STM32F746 MCU (see Fig. 11). Please note that these result plots only consider implementations for primary variants of the candidates. Due to limited space, we do not show further comparisons with non-primary variants or our total result table containing almost 300 different implementations. However, all results and possible updates are available online[4]. Moreover, to always represent the capabilities of each cipher, we do not use only one specific implementation for one cipher for every test case (speed, ROM and RAM). Instead we select the primary implementation with the best result for the metric in question and include this result in the plot. This means that the speed measurement e.g. from Fig. 5 for one cipher might origin from a speed-optimized assembly implementation, while the result for the code size measurement (e.g. from Fig. 9) for the same cipher was obtained by testing a different (size-optimized) implementation of the primary variant. That way we want to acknowledge the fact that different use cases and constraints might ask for differently optimized implementations of LWC algorithms. Please also pay attention to the scaling of the y-axis in the different plots. If the lowest and highest value in the plot differ in orders of magnitude, we choose the standard logarithmic scaling of LaTeX, which makes reading the absolute values a bit difficult. However, it allows for easier qualitative visual inspection of the results.

Since we always include the best test result of each finalist's primary variant for each metric/platform category, we obtain ten result bars. For comparison reasons, we also plot the benchmark result of an AES-GCM implementation.

[4] https://lwc.las3.de.

We stripped the implementation of AES-GCM out of MbedTLS[5], a popular open-source implementation of SSL/TLS, which if often used on embedded systems. Our single implementation of AES-GCM was compiled with the flags MBEDTLS_AES_ROM_TABLES and MBEDTLS_AES_FEWER_TABLES enabled. Adding MBEDTLS_AES_ROM_TABLES to the configuration has the effect that the SBOX/RCON tables of AES are placed in the ROM instead of initializing them in the RAM during runtime. Using MBEDTLS_AES_FEWER_TABLES reduces the amount of precalculated tables which has a positive effect on the binary size. These flags have been set since we believe this configuration brings the AES-GCM implementation closer to most implementations we have seen from the LWC algorithms. Note that the speed/code size benchmarks for the 8-bit AVR ATmega328P chip do not contain any results for AES-GCM. Unfortunately, the implementation does not work on this platform since it is designed and optimized for 32-bit architectures.

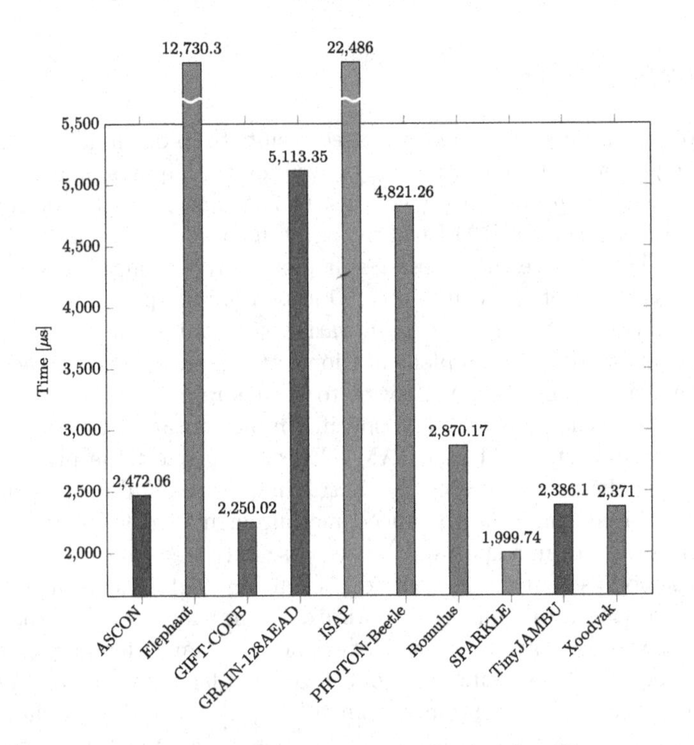

Fig. 1. Speed measurements on the Arduino Uno

The speed benchmark on the AVR platform (see Fig. 1) shows SPARKLE, GIFT-COFB and Xoodyak in the top 3, with ASCON, TinyJAMBU and Romulus being fairly close. GRAIN-128AEAD and PHOTON-Beetle remain in the

[5] https://tls.mbed.org/.

middle group and Elephant and ISAP represent the slowest algorithms in our 8-bit MCU benchmark. We would like to point out that the worse performance of ISAP is somewhat expected since the protection against attacks included in the cipher design leads to a penalty in throughput. Moreover, we recall that we have only a very small amount of implementations available for Elephant and most of them are not specifically optimized. Lastly, SPARKLE and TinyJAMBU offer permutations which can be implemented very efficiently in software which might give them an advantage-by-design in our benchmarks – especially over candidates that are rather designed for high throughput/low area requirements in hardware.

Next, we take a closer look at the throughput numbers for our two most powerful MCUs – the ESP32 and the Sipeed Maixduino (see Figs. 2 and 3). Right away we can see that the bar patterns of the two plots look very similar to each other. ASCON, Xoodyak and TinyJAMBU are leading on both architectures and we can observe similar rankings/patterns also for the other ciphers. However, it is especially interesting that the AES-GCM implementation outperforms five LWC candidates on the ESP32 and ranks even on the 4th spot on the Maixduino.

On the ARM platforms (results depicted in Figs. 4 and 5), we generally find the same LWC candidates in the first spots as on the other platforms. ASCON leads the group in the STM32F7 performance test and is only beaten by Xoodyak and SPARKLE on the STM32F1 MCU. Moreover, while a cipher might move up or down one or two positions on the different architectures, we always find the same LWC ciphers in the upper tier in the speed test case. Namely, Xoodyak, ASCON, GIFT-COFB, SPARKLE and TinyJAMBU form the top half on both ARM MCUs. AES-GCM ranks 6th on the STM32F7 and 8th on the STM32F103. We want to add here that there exist many more ARM-optimized LWC implementations than RISC-V- or Xtensa-optimized implementations, meaning lots of the best results on ARM have been obtained by applying specific assembly implementations, while this was not the case in the ESP32/Maixduino benchmarks. Furthermore, our AES-GCM implementation is optimized for 32-bit systems in general but not solely for the ARM architecture.

In the code size measurements, we can again find similar patterns on different platforms. ASCON, TinyJAMBU, SPARKLE, Xoodyak and ISAP reach the lowest code sizes on the ESP32, Maixduino and STM23F7 (see Figs. 7, 8 and 9). On the STM32F1, GIFT-COFB manages to break this group dominance and ranks on the 3rd place. On AVR architecture, the situation is different. PHOTON-Beetle provides the smallest binary, followed by ASCON, ISAP, Xoodyak and GIFT-COFB. The smallest PHOTON-Beetle implementation is developed specifically for AVR and optimized to have a low ROM footprint. For other candidates no such specialized implementation is available for the AVR architecture. This again shows the emphasis of designers and developers for ARM, for which many more assembly implementations exist. Moreover, we would like to point out that some of the lowest size implementations on any platform are some of the worst in terms of throughput. For a couple of algorithms, the lowest size implementation is their reference implementation, which is a good

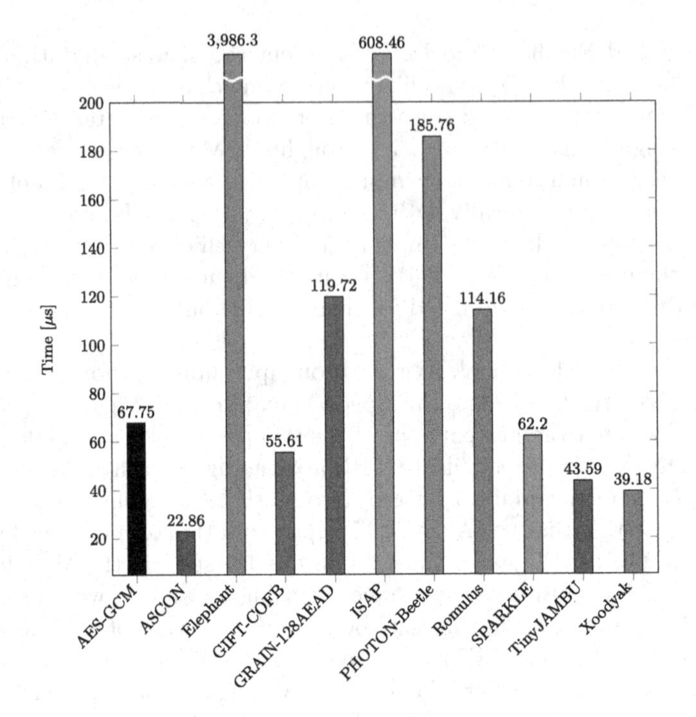

Fig. 2. Speed measurements on the ESP32

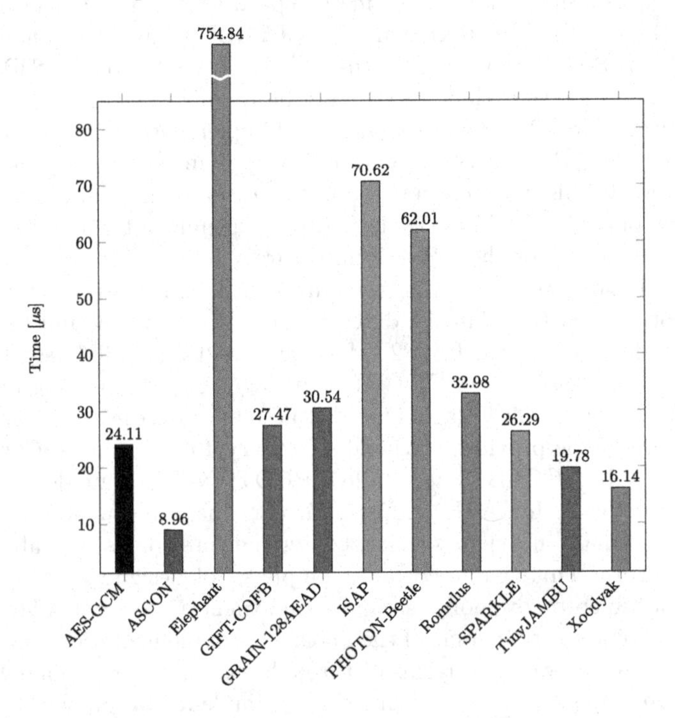

Fig. 3. Speed measurements on the Maixduino

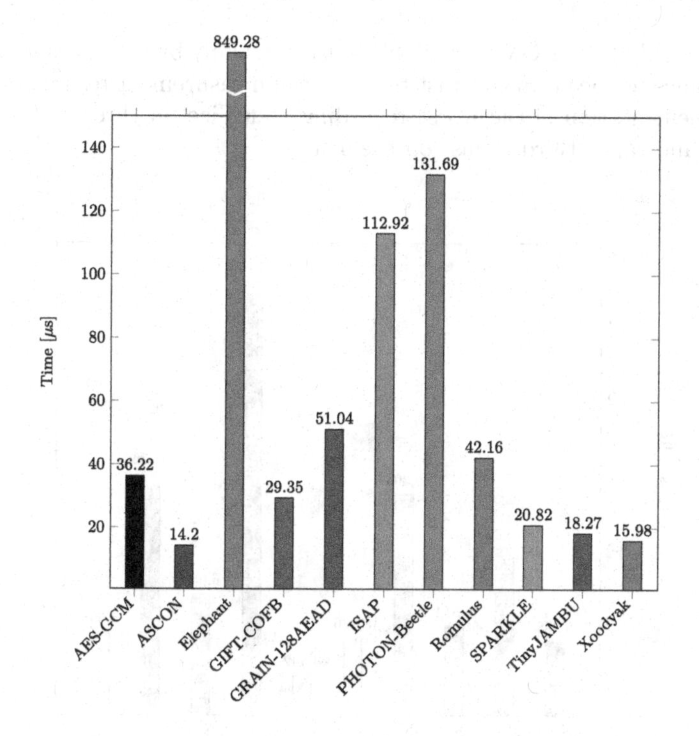

Fig. 4. Speed measurements on the STM32F7

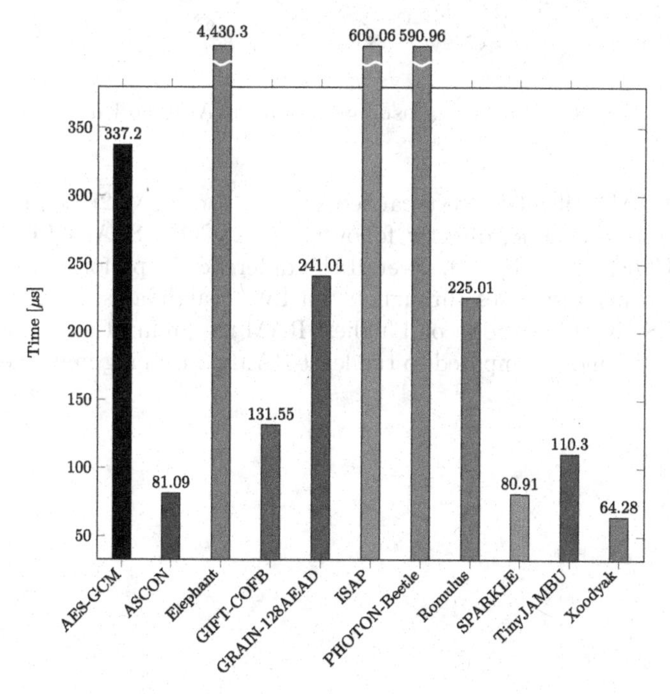

Fig. 5. Speed measurements on the STM32F103

choice when only little ROM is available but it is usually by no means lightweight when it comes to speed. In contrast to the speed measurement results, our AES-GCM implementation is the worst regarding code size on three of four tested platforms and ranks third to last on the 4th.

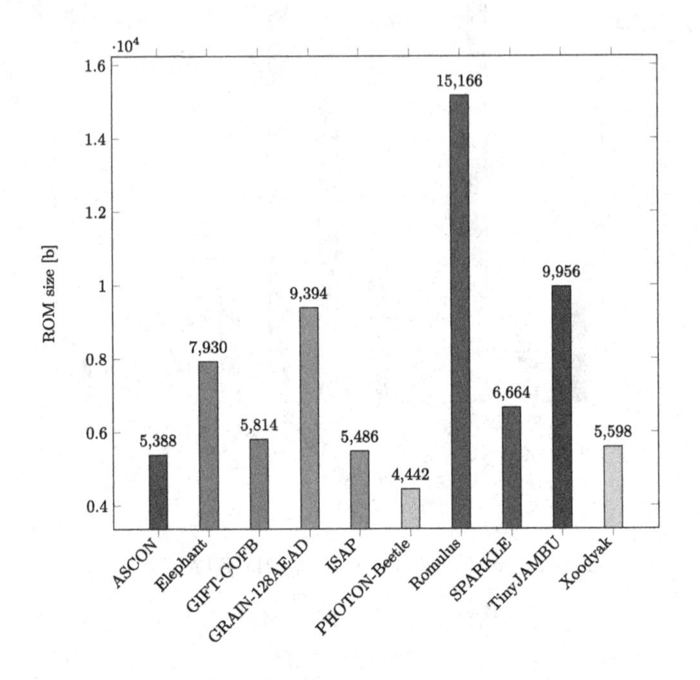

Fig. 6. Code size measurements on the Arduino Uno

In our RAM utilization test carried out on the STM32F7, TinyJAMBU achieves the lowest value, directly followed by ASCON, SPARKLE, Xoodyak and GIFT-COFB (see Fig. 11). Overall the difference in performance is not as severe as in other test cases throughout all LWC candidates. However, in our test case AES-GCM has the second highest RAM requirement among the tested ciphers (ca. 60% more compared to the least RAM-consuming implementation).

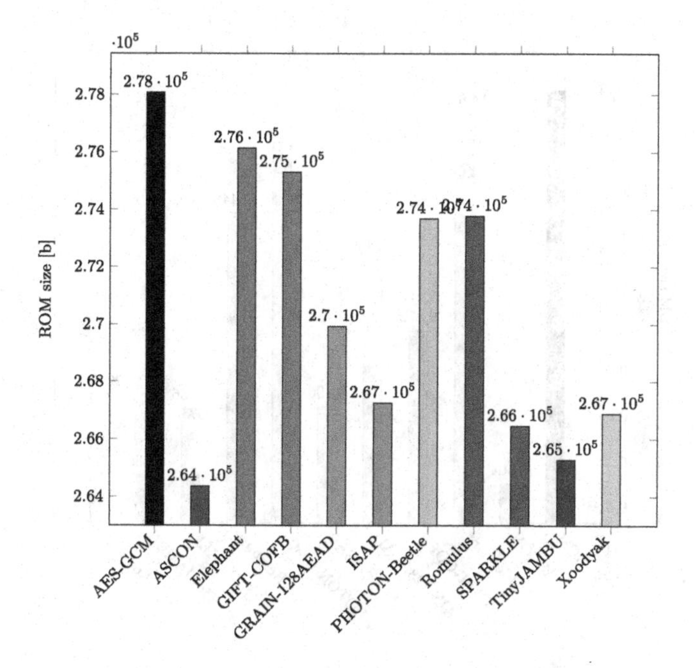

Fig. 7. Code size measurements on the ESP32

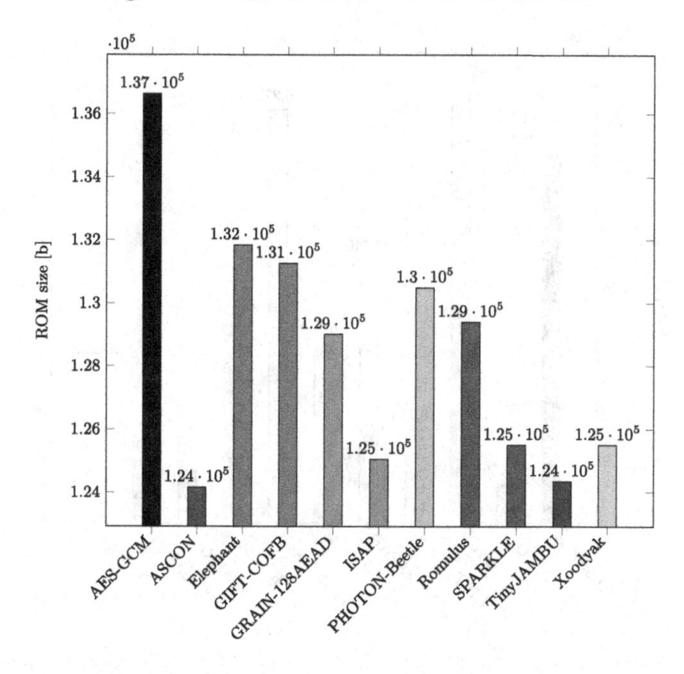

Fig. 8. Code size measurements on the Maixduino

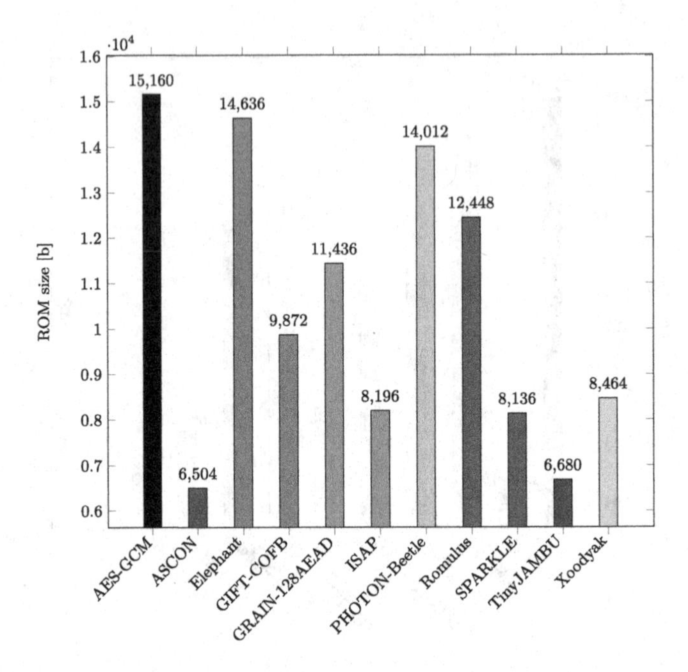

Fig. 9. Code size measurements on the STM32F7

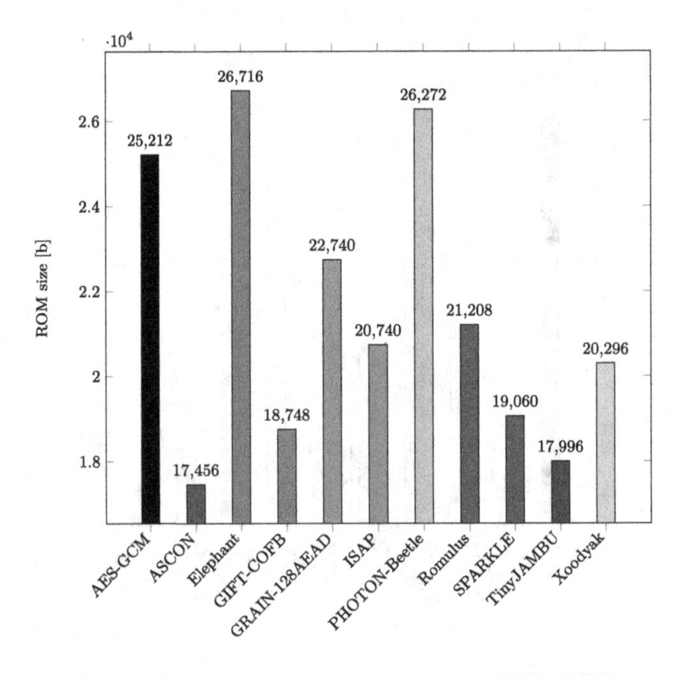

Fig. 10. Code size measurements on the STM32F103

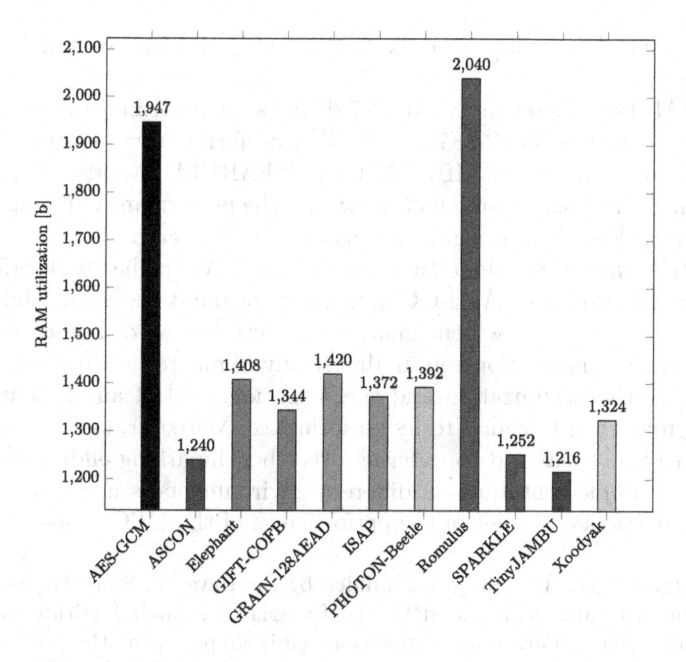

Fig. 11. RAM utilization measurements on the STM32F7

5 Conclusion

We introduced our testing methodology to fairly evaluate the finalists of the NIST LWC project on microcontrollers. We gave some insights on how we benchmark different metrics for lightweight ciphers and provided a description of the candidates' properties. Moreover, we presented performance figures for the best primary implementations of each cipher in various test cases. Our results show that our MbedTLS version of AES-GCM outperforms at least three LWC candidates on each speed benchmark. Depending on the platform, it even reaches up to the 4th rank in the throughput comparison. However, it is always beaten by implementations of Xoodyak, TinyJAMBU and ASCON. Other algorithms that show very good performance in the speed test cases are SPARKLE and GIFT-COFB.

In the code size benchmark the AES-GCM implementation ranks a lot worse than in the speed test case. It allocates the last spot on all but one platform, where it reaches the third to last place. ASCON, SPARKLE, TinyJAMBU and Xoodyak again reach the best overall results on the non-AVR platforms in this test case. GIFT-COFB maintains a lower binary size only on AVR and ARM platforms. ISAP delivers good results in terms of binary size, in contrast to the speed test case. PHOTON-Beetle outperforms all other algorithms regarding code size on the Arduino Uno, where TinyJAMBU can only reach the second to last spot. Since the result of the PHOTON-Beetle was achieved due to lots of architecture-specific optimization of the implementation, it suggests that the right implementation strategy for the target device can lead to massive performance gains.

The RAM test case on the STM32F7 delivers qualitatively similar results to the ROM test on the STM32F7. AES-GCM performs worse than almost any LWC candidate and TinyJAMBU, ASCON, SPARKLE, Xoodyak and GIFT-COFB form the top performing half. However, the performance difference in the RAM benchmark are not as significant as in other test cases.

Ultimately, we can conclude that some of the LWC ciphers outperform our chosen implementation of AES-GCM in every of our test cases, while others achieve that in only very few benchmarks. Furthermore, we can observe that the same algorithms come out on top in almost any of our tests. However, we have also seen that the optimization and specialization level of an implementation plays a big role when in comes to its performance. Moreover, our test setup can only cover certain cases and considering other benchmarking efforts of software and hardware implementations in different environments is necessary, in order to make a holistic assessment of the performance of the LWC candidates.

Acknowledgements. This project is funded by the Bavarian State Ministry of Science and the Arts and coordinated by the Bavarian Research Institute for Digital Transformation (bidt). Furthermore, this research is supported by the BayWISS Consortium Digitization.

References

1. Ankele, R., Ankele, R.: Software Benchmarking of the 2nd round CAESAR Candidates (2016). https://doi.org/10.13140/RG.2.2.28074.26566
2. Atmel Corporation: 8-bit AVR microcontroller with 32k bytes in-system programmable flash. https://ww1.microchip.com/downloads/en/DeviceDoc/Atmel-7810-Automotive-Microcontrollers-ATmega328P_Datasheet.pdf. Accessed 25 July 2022
3. Babbage, S., et al.: The eSTREAM portfolio. eSTREAM, ECRYPT Stream Cipher Project, pp. 1–6 (2008)
4. Banik, S., et al.: GIFT-COFB, Submission to the NIST Lightweight Cryptography Standardization Process (2019). https://csrc.nist.gov/Projects/lightweight-cryptography/finalists. Accessed 25 July 2022
5. Bao, Z., et al.: PHOTON-Beetle Authenticated Encryption and Hash Family, Submission to the NIST Lightweight Cryptography Standardization Process (2019). https://csrc.nist.gov/Projects/lightweight-cryptography/finalists. Accessed 25 July 2022
6. Beierle, C., et al.: Alzette: a 64-bit ARX-box. In: Micciancio, D., Ristenpart, T. (eds.) CRYPTO 2020. LNCS, vol. 12172, pp. 419–448. Springer, Cham (2020). https://doi.org/10.1007/978-3-030-56877-1_15
7. Beierle, C., et al.: Schwaemm and esch: lightweight authenticated encryption and hashing using the sparkle permutation family. In: The NIST Lightweight Cryptography Standardization Process (2019). https://csrc.nist.gov/Projects/lightweight-cryptography/finalists. Accessed 25 July 2022

8. Beierle, C., et al.: The SKINNY family of block ciphers and its low-latency variant MANTIS. In: Robshaw, M., Katz, J. (eds.) CRYPTO 2016. LNCS, vol. 9815, pp. 123–153. Springer, Heidelberg (2016). https://doi.org/10.1007/978-3-662-53008-5_5
9. Bernstein, D.J., Lange, T.: eBACS: ECRYPT benchmarking of cryptographic systems. http://bench.cr.yp.to. Accessed 25 July 2022
10. Berti, F., Guo, C., Pereira, O., Peters, T., Standaert, F.X.: TEDT, a leakage-resist AEAD mode for high physical security applications. IACR Trans. Cryptogr. Hardw. Embed. Syst. pp. 256–320 (2020)
11. Bertoni, G., Daemen, J., Peeters, M., Van Assche, G.: Keccak. In: Johansson, T., Nguyen, P.Q. (eds.) EUROCRYPT 2013. LNCS, vol. 7881, pp. 313–314. Springer, Heidelberg (2013). https://doi.org/10.1007/978-3-642-38348-9_19
12. Bertoni, G., Daemen, J., Peeters, M., Van Assche, G.: Permutation-based encryption, authentication and authenticated encryption. In: Directions in Authenticated Ciphers, pp. 159–170 (2012)
13. Beyne, T., Chen, Y.L., Dobraunig, C., Mennink, B.: Elephant, Submission to the NIST Lightweight Cryptography Standardization Process (2019). https://csrc.nist.gov/Projects/lightweight-cryptography/finalists. Accessed 25 July 2022
14. Bogdanov, A., Knežević, M., Leander, G., Toz, D., Varıcı, K., Verbauwhede, I.: SPONGENT: a lightweight hash function. In: Preneel, B., Takagi, T. (eds.) CHES 2011. LNCS, vol. 6917, pp. 312–325. Springer, Heidelberg (2011). https://doi.org/10.1007/978-3-642-23951-9_21
15. CAESAR committee: CAESAR: Competition for Authenticated Encryption (2019). https://competitions.cr.yp.to/caesar.html. Accessed 25 July 2022
16. Campos, F., Jellema, L., Lemmen, M., Müller, L., Sprenkels, D., Viguier, B.: Assembly or optimized C for lightweight cryptography on RISC-V? In: Krenn, S., Shulman, H., Vaudenay, S. (eds.) CANS 2020. LNCS, vol. 12579, pp. 526–545. Springer, Cham (2020). https://doi.org/10.1007/978-3-030-65411-5_26
17. Cazorla, M., Gourgeon, S., Marquet, K., Minier, M.: Survey and benchmark of lightweight block ciphers for MSP430 16-bit microcontroller. Sec. and Commun. Netw. **8**(18), 3564–3579 (2015). https://doi.org/10.1002/sec.1281
18. Daemen, J., Hoffert, S., Peeters, M., Assche, G.V., Keer, R.V.: Xoodyak: A Lightweight Cryptographic Scheme. In: The NIST Lightweight Cryptography Standardization Process (2019). https://csrc.nist.gov/Projects/lightweight-cryptography/finalists. Accessed 25 July 2022
19. Dinu, D., Biryukov, A., Großschädl, J., Khovratovich, D., Corre, Y.L., Perrin, L.: FELICS - fair evaluation of lightweight cryptographic systems. In: NIST Workshop on Lightweight Cryptography (2015)
20. Dinu, D., Corre, Y.L., Khovratovich, D., Perrin, L., Großschädl, J., Biryukov, A.: Triathlon of lightweight block ciphers for the Internet of things. J. Cryptogr. Eng. **9**(3), 283–302 (2018). https://doi.org/10.1007/s13389-018-0193-x
21. Dobraunig, C., et al.: ISAP v2.0. In: The NIST Lightweight Cryptography Standardization Process (2019). https://csrc.nist.gov/Projects/lightweight-cryptography/finalists. Accessed 25 July 2022
22. Dobraunig, C., Eichlseder, M., Mendel, F., Schläffer, M.: Ascon v1.2. In: The NIST Lightweight Cryptography Standardization Process (2019). https://csrc.nist.gov/Projects/lightweight-cryptography/finalists. Accessed 25 July 2022
23. Espressif Systems: ESP32WROOM32 Datasheet. https://www.espressif.com/sites/default/files/documentation/esp32-wroom-32e_esp32-wroom-32ue_datasheet_en.pdf. Accessed 25 July 2022

24. Guo, J., Peyrin, T., Poschmann, A.: The PHOTON family of lightweight hash functions. In: Rogaway, P. (ed.) CRYPTO 2011. LNCS, vol. 6841, pp. 222–239. Springer, Heidelberg (2011). https://doi.org/10.1007/978-3-642-22792-9_13
25. Hell, M., Johansson, T., Maximov, A., Willi Meier, F., Sönnerup, S.J., Yoshida, H.: Grain-128AEADv2- A lightweight AEAD stream cipher. In: The NIST Lightweight Cryptography Standardization Process (2019). https://csrc.nist.gov/Projects/lightweight-cryptography/finalists. Accessed 25 July 2022
26. Hyncica, O., Kucera, P., Honzik, P., Fiedler, P.: Performance evaluation of symmetric cryptography in embedded systems. In: Proceedings of the 6th IEEE International Conference on Intelligent Data Acquisition and Advanced Computing Systems, vol. 1, pp. 277–282, September 2011. https://doi.org/10.1109/IDAACS.2011.6072756
27. Iwata, T., Khairallah, M., Minematsu, K., Peyrin, T., Guo, C.: Romulus v1.2. In: NIST Lightweight Cryptography Standardization Process (2019). https://csrc.nist.gov/Projects/lightweight-cryptography/finalists. Accessed 25 July 2022
28. NIST LWC team: Microcontroller Benchmarking (2021). https://github.com/usnistgov/Lightweight-Cryptography-Benchmarking/. Accessed 25 July 2022
29. Renner, S., Pozzobon, E., Mottok, J.: A hardware in the loop benchmark suite to evaluate NIST LWC ciphers on microcontrollers. In: Meng, W., Gollmann, D., Jensen, C.D., Zhou, J. (eds.) ICICS 2020. LNCS, vol. 12282, pp. 495–509. Springer, Cham (2020). https://doi.org/10.1007/978-3-030-61078-4_28
30. Cardoso dos Santos, L., Großschädl, J., Biryukov, A.: FELICS-AEAD: benchmarking of lightweight authenticated encryption algorithms. In: Belaïd, S., Güneysu, T. (eds.) CARDIS 2019. LNCS, vol. 11833, pp. 216–233. Springer, Cham (2020). https://doi.org/10.1007/978-3-030-42068-0_13
31. Seeed Studio: Sipeed Maixduino Specifications v1.0. https://www.mouser.de/pdfDocs/SipeedMaixduinoSpecifications_ENV10.pdf. Accessed 25 July 2022
32. for Standards, N.I., Technology: Submission requirements and evaluation criteria for the lightweight cryptography standardization process (2018). https://csrc.nist.gov/CSRC/media/Projects/Lightweight-Cryptography/documents/final-lwc-submission-requirements-august2018.pdf. Accessed 25 July 2022
33. STMicroelectronics: STM32F103x8 Datasheet. https://www.st.com/resource/en/datasheet/stm32f103c8.pdf. Accessed 25 July 2022
34. STMicroelectronics: STM32F746xx Datasheet. https://www.st.com/resource/en/datasheet/stm32f746ng.pdf. Accessed 25 July 2022
35. Tschofenig, H., Pegourie-Gonnard, M.: Performance of state-of-the-art cryptography on arm-based microprocessors. In: NIST Workshop on Lightweight Cryptography (2015)
36. Weatherley, R.: Lightweight Cryptography Primitives (2021). https://rweather.github.io/lightweight-crypto/performance.html. Accessed 25 July 2022
37. Wu, H., Huang, T.: JAMBU lightweight authenticated encryption mode and AES-JAMBU. In: CAESAR Competition Proposal (2014)
38. Wu, H., Huang, T.: TinyJAMBU: a family of lightweight authenticated encryption algorithms. In: The NIST Lightweight Cryptography Standardization Process (2019). https://csrc.nist.gov/Projects/lightweight-cryptography/finalists. Accessed 25 July 2022

Evolving a Boolean Masked Adder Using Neuroevolution

Sebastian Renner[1,2(✉)], Enrico Pozzobon[1,3], and Jürgen Mottok[1]

[1] OTH Regensburg, Regensburg, Germany
{sebastian1.renner,enrico.pozzobon,juergen.mottok}@othr.de
[2] Technical University of Munich, Munich, Germany
[3] University of West Bohemia, Pilsen, Czech Republic

Abstract. The modular addition is a popular building block when designing lightweight ciphers. While algorithms mainly based on the addition can reach very high performance, masking their implementations results in a huge penalty. Since efficient protection against side-channel attacks is a requirement in lots of use cases, we focus on optimizing the Boolean masking of the modular addition. Contrary to recent related work, we target evolving a masked full adder instead of parts of a parallel prefix adder. We study how techniques typically found in neural network evolution and genetic algorithms can be adapted in order to help in evolving an efficiently masked adder. We customize a well-known neuroevolution algorithm, develop an optimized masked adder with our new approach and implement the ChaCha20 cipher on an ARM Cortex-M3 controller. We compare the performance of the protected neuroevolved implementation to solutions found by traditional search methods. Moreover, the leakage of our new solution is validated by a t-test conducted with a leakage simulator. We present under which circumstances our masked implementation outperforms related work and prove the feasibility of successfully using neuroevolution when searching for complex Boolean networks.

Keywords: Neuroevolution · Modular addition · Masking · ChaCha20 · Side-channel analysis

1 Introduction

Since the number of low-power IoT devices constantly increases and e.g. intelligent sensors are also used in critical environments, there is a rising demand for data protection through less resource-intensive cryptography. This is also reflected by the lightweight cryptography (LWC) project initiated by the Standards and Technology (NIST). The NIST LWC project was started in 2018 and aims to validate cipher submissions for their use in constrained environments. Currently, the competition reached the 3rd and final evaluation round which is expected to yield at least one cipher that gets standardized in a Federal Information Processing Standard (FIPS) in the following. While a small memory

© The Author(s), under exclusive license to Springer Nature Switzerland AG 2022
W. Li et al. (Eds.): ADIoT 2022, LNCS 13745, pp. 21–40, 2022.
https://doi.org/10.1007/978-3-031-21311-3_2

footprint and high throughout are obvious requirements for LWC algorithms, they also need to provide a certain security level depending on the use case. Often, high performance and a higher resistance against attacks are contradicting requirements. Some ciphers might be operated at high speeds when implemented without additional security measurements but experience a big performance penalty if additional protection is needed. Therefore, efficient protection against e.g. basic side-channel attacks poses an interesting field of research in the domain of (lightweight) cryptography.

A popular design pattern for LWC is to compose an algorithm only out of modular additions, rotations and exclusive-or (XOR) operations. Speck, ChaCha20 and Sparkle – which is a 3rd round candidate in NIST's LWC project – are examples for such ARX-based ciphers. The simple structure of these algorithms allows for very fast implementations. However, the protection against certain side-channel attacks such as the bricklayer attack on ChaCha20 results in a high performance penalty [1]. This stands out even more, when compared to masked representations of other ciphers, e.g. AES, where the overhead is not as severe [2,19]. There exists a number of publications which suggest different approaches to mitigate this disadvantage in order to make ARX ciphers more usable in protected scenarios [3,9,14].

1.1 Contributions

In this paper, we introduce a variant of a neuroevolution algorithm which operates on Boolean networks. We explain our changes made to the original implementation and use our customized version to search for an efficiently masked adder needed in protected variants of ARX ciphers. We select this use case to evaluate the feasibility of incorporating neuroevolution to find solutions to such Boolean problems. We show that it is possible to evolve fitting results with our alternative approach. We compare the performance of our neuroevolution search to state of the art solutions. We implement the ChaCha20 ARX cipher including the different adder variants and evaluate their performance on a STM32F103 microcontroller. To verify the correctness of the leakage detection mechanism within our genetic algorithm, we simulate the power consumption during execution and conduct a t-test with MAPS [7].

1.2 Paper Organization

In Sect. 2, we introduce related approaches to mask the modular addition, before explaining our problem setup and optimizing strategy incorporating neuroevolution in detail. In Sect. 3, we present our results obtained through customized neuroevolution techniques. In Sect. 4, we elaborate on how we implemented our solutions in ARM assembly and present benchmark results obtained on an ARM Cortex-M3 MCU. We compare different masked adders within an implementation of ChaCha20 and report the performance of related and our own work for various message sizes. Section 5 briefly explains how we verified the absence of distance-based leakage using the simulator MAPS. Section 6 concludes the paper and discusses possible future work.

2 Efficient Side-Channel Protection

In this work, we investigate how masking the modular addition operation can be further optimized. We choose the number of shares d to be 2, which leads to protection against first order Differential Power Analysis (DPA) attacks, since d share implementations prevent attacks of an order of up to $d - 1$. In practice, this means that an attacker probing one wire on a device running a two-share implementation cannot extract sensitive information caused by distance-based leakage. In a real-world scenario, possible leakage originating from the chip's pipeline and registers needs to be taken into account and additional protection measures need to be put into place. In the following, we will present our optimized adder implementation protected against distance-based leakage. This implementation is most suited to be compared to related work, since e.g. the Threshold Implementation (TI) adder published by Jungk et al. does also not take pipeline-based leakages into account [14].

2.1 Conventional Masking of Addition

Research on efficiently masking modular addition dates back to the early 2000s. Older works suggest a mixture of Boolean and arithmetic masking depending on the type of operation [5]. This approach requires routines to convert Boolean into arithmetic masks and vice versa. Especially, the conversion from arithmetic to Boolean leads to a high performance overhead. Goubin's proposed algorithm has a complexity of $O(k)$ – with k being the bit width of the addition – while later work by Coron et al. showed that can be reduced to $O(\log k)$ [6,11]. Authors of recent work on masking the addition stepped away from the idea of optimizing the conversion between the different types of masking. For example, Biryukov et al. focused on a direct use of Boolean masking for all operations instead of blending Boolean and arithmetic masking together [3]. Their work is based on using the Kogge-Stone adder (KSA) and masking its individual gates. They present an exhaustive-search algorithm to find the optimal masked representation of the various gates needed in the KSA. The verification of the masking of the sensitive values is done with a t-test. The secure KSA is then built from the optimal secure gates. This approach leads to a significant performance gain over the earlier work from Coron et al. Other research that targets Hardware instead of software implementations was conducted by Schneider et al. They presented a TI of a masked addition featuring three shares [18]. While their solution is delivering high performance in a Hardware setting, it is not optimized for software implementations. Jungk et al. picked up the TI approach and showed an application to the software domain [14]. Similar to Biryukov et al., they base their work on the KSA and search for efficiently masked representations of different gates. Moreover, Jungk et al. do not focus on elementary structures like XOR, AND or SHIFT only, but also provide secure (masked) gadgets, meaning useful combinations of certain operations (e.g. SecAndShift). They present a two-share TI for each of the used gates/gadgets in the KSA and assemble the adder out of these structures. Through their TI approach applied to software implementations, they can achieve a performance gain over other previously published masked adders. This is shown in a benchmarking experiment where Jungk et al. compare their optimized version of ChaCha20 to other published implementations of the cipher.

In this work, we do not focus on masking individual parts of a parallel prefix adder. Instead, we search for an efficiently masked full adder (with inputs a, b, c_{in} and outputs s and c_{out}) which can later be used in bitsliced software implementations. We choose this adder structure above e.g. a KSA since a) the problem size of a 32-bit adder would exhaust the capabilities of our neuroevolution algorithm and b) optimizing specific gates and adder gadgets has already been researched and improved to its maximum. Through customizing and engaging neuroevolution, we would like to introduce a novel way of tackling such complex Boolean problems. Similar to the research of Biryukov et al., we do not manually design a possible solution. While they operate on a smaller problem size which allows for standard exhaustive-search, we evaluate if a more open search process based on neuroevolution and genetic algorithms yields equivalent or better results. Our goal is to show an alternative path when dealing with Boolean problems similar to the masked adder use case.

2.2 Optimizing Masked Addition Using Neuroevolution

Neuroevolution, a term originating from the domain of artificial intelligence, describes the use of genetic algorithms for generating and evolving (artificial) neural networks [10,23]. In contrast to gradient-descent-based approaches, neuroevolution builds upon genetic algorithms, imitating evolving processes known from nature. Neuroevolution techniques are used to solve reinforcement learning problems. Often, the capabilities of these techniques are showcased in their application to video games [13,22]. Generally, neuroevolution methods can be separated into conventional and Topology and Weight Evolving Artificial Neural Network (TWEANN) algorithms. While the former only alter connection weights of neural networks, the latter also change the topology, i.e. the connections and nodes themselves during evolution. The underlying genetic algorithm follows the same general phases in different neuroevolution techniques: initialization, selection, crossover and mutation. Always a population p_0 consisting of n networks is first initialized according to the configuration of the algorithm. A previously defined fitness function, which describes the desired behavior of the search network, then evaluates all n members of the population. A certain percentage of the best-fit candidates is selected and proceeds to the crossover phase. Here, networks are mated to produce the next generation. Also, some members of the new generation encounter mutation, similar to what happens in nature. After this last phase is completed, the algorithm has finished one evolution cycle and now performs the same process again, but now on the child generation p_1. Depending on the success of the computations, the algorithm is either carried out for m evolution cycles or terminates after a perfectly fitting solution (network) for the given problem has been found.

A popular neuroevolution method based on the just described cycle is the NeuroEvolution of Augmenting Topologies (NEAT) algorithm, which is operating on TWEANNs. It was introduced by Stanley et al. in 2002 and it has been shown that it outperforms other (neuroevolution) techniques for various tasks [4,16,20,21]. This reputation in the artificial intelligence domain, well documented open-source implementations and its possibility to work on a network's topology are the reasons why we chose NEAT as our baseline algorithm, which we customized to work on Boolean networks.

Despite of the implementation of a standard genetic algorithm, NEAT offers a few specialities which differentiate it from other methods in this field of research. Genomes are an encoding of networks. A genome (network) consists of both nodes (node genes) and connections (connection genes). Each node gene can be connected to other node genes through various connection genes. Depending on where the node gene is located within the network, it is referred to as an input, a hidden or an output node. Each connection corresponds to an input node, an output node and has a weight. During mutation, a connection's weight can be altered or the status of the connection itself can be set to be en- or disabled. Moreover, additional connection genes can be added into the genome. Node genes can also be inserted into the network. In that case, an existing connection is split and a node gene is initialized at the breakpoint. The old connection is disabled and the new node is integrated by receiving two new connections.

To make use of these biology-inspired features of NEAT and able to apply a genetic selection process on Boolean networks, we needed to make various adjustments to the design of NEAT. More precisely, we had to take an implementation of the NEAT algorithm and modify it in order to support Boolean problems. Usually, NEAT is used to evolve neural networks which solve continuous problems. Similar to other neural network algorithms, NEAT is operating on different properties of a network and its genes (nodes and connections) to evolve an optimal solution to its given problem. The properties like biases, responses, weights, aggregation and activation functions are implemented in a continuous setting, meaning they represent or work on floating point values. These values and functions are used to determine the output of a gene or the whole network [20]. The output of a node gene is calculated as seen in Eq. 1.

$$output = activation(bias + (response * aggregation(inputs))) \quad (1)$$

Since obviously floating point parameters are not appropriate for our masked adder representation, we customized the NEAT implementation such that it is applicable to this kind of Boolean and discrete problems. First off, we removed all the unnecessary properties that we do not need for our experiment setting. This includes weights, biases, responses and activation functions. Our goal in this step was to reduce the properties to those that are essentially needed for solving Boolean problems. The general idea we followed was to let NEAT work on Boolean networks, in which the node genes would represent atomic logical expressions. Connection genes should only feed input values into the custom node genes. These inputs could originate from the initial input to the network or consist of intermediate values that had already been processed by other nodes. We limited the weight property of the connection genes to the values 0 and 1, essentially resulting in two states: Either a connection is enabled and feeds values to the connected nodes or it is (temporarily) disabled and can therefor not transport values to following nodes. Our node genes only implement an aggregation function, which is directly applied to its inputs. This simplifies the calculation of a node's output to Eq. 2.

$$output = aggregation(inputs) \quad (2)$$

We implemented and enabled only custom aggregation functions representing simple logic gates. This way we make sure that we can realize each node aggregation in one ARM assembly instruction, either directly or after a conversion. The available aggregations are XOR, OR, AND, NOR, NAND and NOT. These modifications allow us to use NEAT to evolve Boolean networks. Note that we did not change the underlying NEAT algorithm in this first step. However, we altered the properties of the (neural) networks that NEAT's genetic algorithm is working on.

To test the Boolean environment of the modified implementation, we need to define the problem that we would like to solve with the help of NEAT. We chose an efficient masked full adder using two shares per in- and output as our design goal. This gives us the advantage that we can run NEAT on a problem with fairly manageable complexity while we can still observe the full adder as a whole, without needing to separate it in into individual parts. As the standard full adder is using single bits for each in- and output, we can later use a potential solution in a bitsliced software implementation of the modular addition of an ARX cipher. We define the inputs of the full adder as a_0, a_1, b_0, b_1, c_{in0} and c_{in1} and the outputs as s_0, s_1, c_{out0} and c_{out1} with v_0 and v_1 representing the two shares of a variable v. On the NEAT side, we set the names and numbers of variables and integrate the truth table for the full adder. This enables NEAT to check the output for each of the 2^6 possible inputs on all potential adder networks which can then be used to determine the adder fitness (correctness) of a particular network.

By default, the NEAT software we used, neat-python, implements one scalar fitness value which is an attribute of each genome (network). The fitness evaluation is conducted for each member of the current population and once in every generation. Moreover, a genome's fitness is the main indicator when the algorithm decides if the member is allowed to reproduce for the subsequent generation. The function that realizes the calculation of the fitness is specific to the individual problem and needs to be provided during setup. The user can also configure a fitness threshold that is only reached when the problem is solved optimally according to the fitness function [15]. We decided to set our fitness threshold to 0. In the evaluation of the networks we compare the output of each genome to the actual truth table of the full adder. Since our adder is operating with two shares per output, we evaluate if $c_{out0} \oplus c_{out1}$ equals c_{out} and $s_0 \oplus s_1$ equals s for every of the 2^6 possible inputs. We set an initial adder fitness value of 0 for each genome and subtract 1 for every wrong output value. Taking into account 64 different inputs with two output values each, the minimal adder fitness is -128, while the fitness goal is 0 – which is only reached when the network represents a fully logically correct adder. With this fitness function setup tied together with the Boolean aggregations, we can already evolve a full shared adder. However, in this development stage potential leakage of secret values is not yet considered and thus any solution will likely be insecure.

We introduced a second fitness value, the leakage fitness, to also take distance-based leakage of the (adder) network into account. In order to evaluate the leakage of a candidate, we implement a leakage check algorithm similar to the one layed out by Gross et al. [12]. We provide a table consisting of a row for every possible combination of (shared) inputs. We then group the shared inputs by secret inputs, e.g. the two input vectors where $a_0 = a_1 = b_0 = b_1 = c_{in0} = c_{in1} = 1$ and $a_0 = a_1 = b_0 = b_1 = c_{in0} = c_{in1} = 0$ both correspond to the same secret input vector $a = b = c_{in} = 0$. We then check which output the network calculates for each intermediate value. We calculate the output (values) at each node of the network. When the Hamming weight (HW) for all secret input groups is the same at a node, we can derive that no distance-based first-order leakage occurs at that point of the network. This is because the equality of the HW corresponds to a statistical independence of the intermediate value and the secret inputs. We set our leakage fitness threshold also to the value 0 and subtract from it in case we detect leakage on any node. In the full adder setting we have 2^3 secret input combinations, meaning we need to calculate 8 HWs for every intermediate node in each network. We subtract 1 for every unequal HW at an intermediate point. This means the leakage penalty for one intermediate value could at most be 8, assuming 8 different HWs. The minimal leakage fitness can be written as $8 * n$ with n being the number of intermediate value nodes. Since we subtract 1 for every unequal HW at a node, no first-order leakage would result in zero penalization due to perfect HW distribution. A HW table with an evenly distributed HW at all 16 nodes is shown in Fig. 1. Note that each column t_n in the table shows the HW distribution at (the output of) one node in the network. According to our fitness function, one unequal value in one column would already show there is leakage in the network and the leakage fitness would be set to -1. Pairing the leakage evaluation with the adder fitness ensures the desired and correct behavior of the network, such that we can conclude that our first-order secure shared full adder has to have a fitness vector of $(0, 0)$, each 0 representing the adder/leakage fitness of a network.

With the two fitness goals, adder correctness and first-order security, we essentially deal with a multi-objective optimization (MOO) problem. It becomes an important question how to determine which network should be allowed to reproduce or survive the selection process and become a member of the subsequent generation. In a single-objective setting, the population is sorted by fitness and a survival threshold dictates which top percentage of the population is transferred to the next algorithm iteration through reproduction. Since the networks should be optimized towards two objectives, we need to experiment with different

Secret Inputs	t_0	t_1	t_2	t_3	t_4	t_5	t_6	t_7	t_8	t_9	t_{10}	t_{11}	t_{12}	t_{13}	t_{14}	t_{15}
0, 0, 0	4	6	4	4	6	4	4	4	4	4	6	6	4	4	4	4
0, 0, 1	4	6	4	4	6	4	4	4	4	4	6	6	4	4	4	4
0, 1, 0	4	6	4	4	6	4	4	4	4	4	6	6	4	4	4	4
0, 1, 1	4	6	4	4	6	4	4	4	4	4	6	6	4	4	4	4
1, 0, 0	4	6	4	4	6	4	4	4	4	4	6	6	4	4	4	4
1, 0, 1	4	6	4	4	6	4	4	4	4	4	6	6	4	4	4	4
1, 1, 0	4	6	4	4	6	4	4	4	4	4	6	6	4	4	4	4
1, 1, 1	4	6	4	4	6	4	4	4	4	4	6	6	4	4	4	4

Fig. 1. Each cell represents the sum of the HWs of the output of one gate (column) grouped by the same secret input (row)

reproduction strategies. We applied various classical MOO selections methods, mainly a weighted sum approach and the nondominated sorting genentic algorithm II (NSGA-II) [8,24]. However, we recognized that the fight between our two fitness goals leads to a stagnation of the evolution algorithm. That is why we then incorporated a completely different selection strategy – novelty search.

In contrast to standard fitness-based reproduction ideas, the novelty search method follows another principle. In the corresponding paper, Risi et al. propose to abandon the candidate's fitness completely during the selection and mating of suitable next generation members [17]. Instead, they suggest to measure the novelty of each candidate and reward the most novel networks by allowing them to reproduce. The way novelty is measured is dependent on how novel behavior can be described for the individual problem. Risi et al. apply NEAT paired with novelty search to a pathfinding problem, in which an agent has to navigate through a maze to reach the target position without crashing into a wall. In this setting, a candidate taking a different route than its average competitor would be rewarded with a higher novelty score, despite it might finish farther away from the target coordinates. Figure 2 illustrates the novelty approach with showing agents navigating through a maze. The pink rectangle marks the start of each agent, the blue one marks the desired target. An agent should optimally find the target position without touching one of the walls. The circles represent agent candidates and where they crashed into a maze wall. When classic fitness selection is used, an algorithm would prefer the closest agents for reproduction (filled with gray), novelty search would allow candidates taking new and unknown routes to reproduce into the next round (see the yellow-filled circles). This small example shows that preferring novel solution paths can aid in tackling deceptive problems. The authors argue that novelty search can be more efficient than fitness-based approaches when solving hard problems. Especially, in experiments where the fitness landscape is not continuous and a small change in the genome could lead to a high improvement in fitness, novelty search is to be considered as an alternative reproduction method.

However, while defining what novelty actually is seems to be straightforward in the maze example, how to measure novelty when operating on Boolean logic

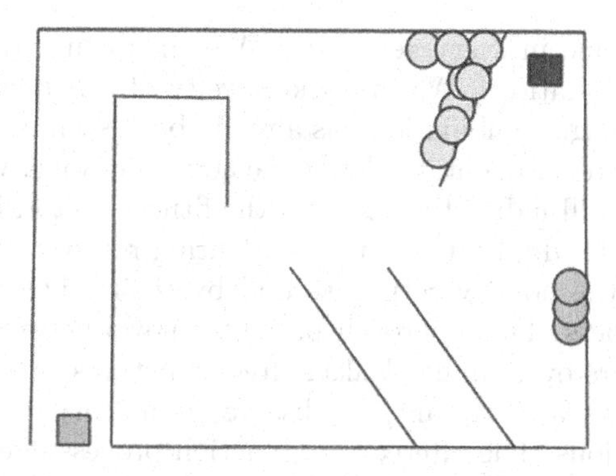

Fig. 2. Target-seeking agents navigating through a maze structure

networks poses a difficult challenge. Since we do not have a fixed target and paths to reach that, we cannot easily evaluate a network when abandoning its fitness completely. Our only indicators for justifying the suitability of a network are its outputs and its HW distributions. With that information we can still record how often a specific fitness vector has been present in past iterations. More precisely, we can introduce an output vector attribute for each network, in which one output tuple (s, c_{out}) is saved for one individual input stimulation. We can then count the occurrences of every appearing output/leakage combination and reward combinations which have rarely or never been seen throughout the whole NEAT run and all past generations. Note that we define our novelty as a combination of a network's outputs and its leakage fitness. With that hybrid approach we still select networks based on one part of their fitness vector (the leakage), however, we completely disregard the adder fitness in the calculation of the novelty because we only rate the networks on the appearance frequency of their outputs.

3 Results

We implemented and experimented with all of the MOO techniques mentioned in the previous section. However, more naive approaches like a weighted sum fitness or Pareto-based selection variants such as NSGA-II did not yield a leakage-free adder structure. We either ended up with a non-secure adder, or a protected gate structure that did not fulfill correctness. Due to the attribute that a small change in our Boolean network can lead to a big jump in fitness, we decided to take a step away from basing the reproduction of genomes only on their fitness. If we consider a change of the aggregation function of one node from an OR to an AND during mutation, it is clear that such a subtle change can have very high impact on a network's fitness. The same assumption holds true for a deletion or rerouting of only one input connection of an arbitrary node. To respect the lack of continuity of the fitness evaluation, we introduced a variant of novelty search into our reproduction routine. We observe the list of outputs and the leakage fitness of every genome in every generation. We then count the occurrences of output/leakage combinations. We rate the novelty of a network based on how often its output/leakage combination has already been seen in past generations. The higher the occurrence count, the lower the novelty reward. While this variant of novelty search is still indirectly based on the fitness vector, it can still detect and disregard logically duplicated networks. During reproduction, the genomes are sorted first by the novelty rating, second by their adder fitness and third by their leakage fitness. This approach promotes novel network structures and prefers correct adders over similar leakage-free structures. After the implementation of our novelty search variant, we observe an improvement in the category of best overall solutions. The altered reproduction process does not evolve perfect solutions, however we can produce networks with a (adder, leakage) fitness vector of (0, −4). These genomes represent full adders with minimal leakage. In fact, we can often evolve an adder which leaks only at the output of a single intermediate node.

With the solution evolved with novelty search being so close to the desired output, we decided to apply a second error-correcting NEAT run in order to obtain a fitness vector of (0, 0). Because the leakage penalty originates from only one node in the network in our best solutions, the second run should specifically optimize that network joint regarding leakage without altering the already satisfactory logic. We found that the small amount of leakage in one node only contributed to one output share, specifically to either c_{out0} or c_{out1}. That showed that our (0, −4) NEAT generated network was already meeting our requirements for 3 of 4 output shares. Knowing that, we can set up a second problem where we only consider the erroneous share a an output. Since an adder fitness of 0 means that the logical output of that share is already correct, we can derive the desired output values from the leaking (0, −4) genome. We then let NEAT evolve a solution using all 6 input shares but only one output share. Again, the goal here is to reach a fitness vector of (0, 0). This second stage problem is easier to solve since the solution does not need to generate 4 correct and non-leaking outputs, but only one. Besides the different output conditions, the configuration of the second NEAT run is equivalent to the first. We also use novelty search for selection and reproduction and do not change any other parameters like e.g. the mutation rate. Due to the lower complexity of the problem, we are able to reach a fitness of (0, 0) in the second NEAT stage. We can then use the suiting network from the second run and replace the leaking network path for c_{outn} with this non-leaking logical twin. This proves that we can evolve a shared first-order-secure full adder using our variant of two-stage novelty search in the custom Boolean NEAT implementation. Figures 3 and 4 show the output of the first NEAT run and the patched network including the result of the second NEAT run. Figure 3 represents a correct adder, however the structure leaks at and only at node 10 which is part of the path or output at the share c_{out1}. Figure 4 includes the leakage-free path for share c_{out1} that we evolved in the second NEAT stage. This network is still a correct full adder while it is free from distance-based leakage according to the definition in Sect. 2.2.

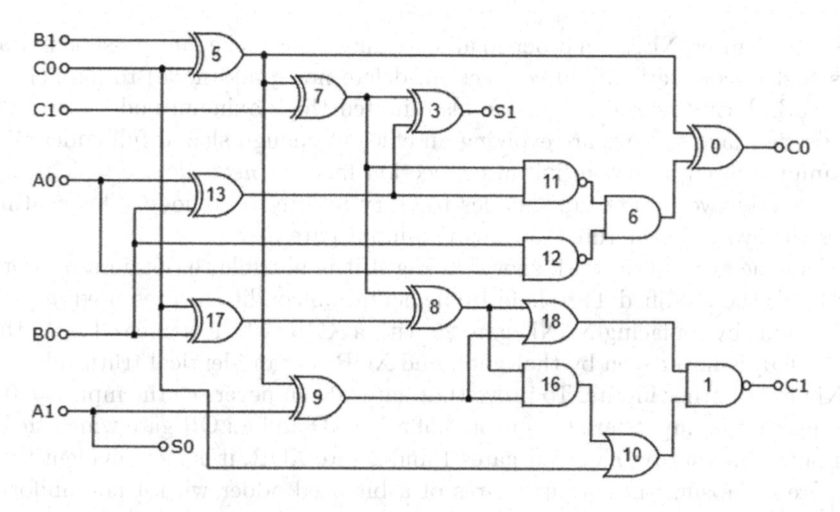

Fig. 3. Shared full adder with distance-based leakage at node 10

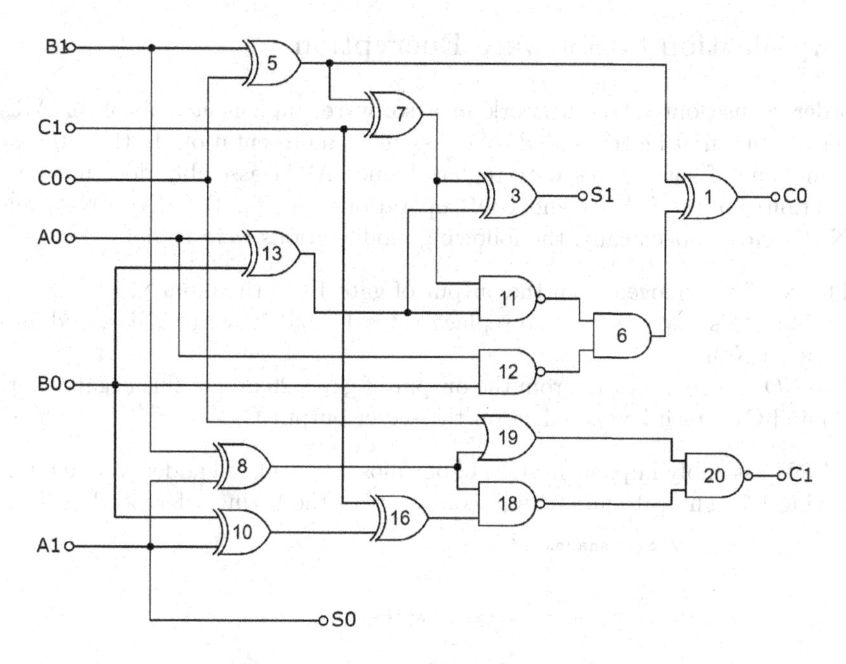

Fig. 4. First-order leakage-free shared full adder network

We configured NEAT in order to allow at most one of the four possible mutations (delete connection, add connection, delete node, add node) to happen to a network during reproduction. We also limited the maximum node size to 20 in order to make sure we are evolving an efficient enough shared full adder. We also implemented a custom initial connection method *neat_double* which connects exactly two inputs/input nodes to every intermediate node. This routine biases the evolved structures towards two-input gate nets.

From the evolved network shown in Fig. 4 it is possible to obtain a network that fulfils the modified Threshold Implementation conditions presented in [14]. This is done by replacing NAND gate 20 with a XOR, which is allowed since the input (0, 0) is never seen by that gate, and XOR has an identical truth table to NAND for all other inputs. To prove that gate 20 can never see the input (0, 0), consider that its inputs are the outputs of a NAND and an OR gate which share an input. Now that both output gates 1 and 20 are XOR, it is more evident that they are collapsing the output shares of a bitsliced adder with 4 non-uniform output shares into 2 uniform output shares.

4 Application to Software Encryption

In order to use our adder network in a software implementation of an ARX cipher, it must first be translated to an assembly representation. In this process, the functions of some gates were replaced since ARM assembly does not have instructions for the NAND and NOR operations, but for BIC (And Not) and ORN (Or Not). Specifically, the following modifications were made:

- The NOT was moved from the output of gate 18 to the input of gate 20.
- De Morgan's law was used to replace gates 11 and 12 with ANDs, and gate 6 with a NOR.
- The NOTs were removed from the output of gates 20 and 6; this negates both C0 and C1 which has no effect on the secret output C.

The ARM assembly implementation of our most efficient full adder can be found in Listing 1.1. The gate numbers in comments of the listing reference Fig. 4.

```
// r6, r7 are shares of A
// r8, r9 are shares of B
// r4, r5 are shares of C
// r0, r1 are used as scratch space
    eor    r2, r6, r8  // gate 13
    and    r6, r8      // gate 12
    eor    r8, r7      // gate 10
    eor    r0, r7, r9  // gate 8
    eor    r8, r5      // gate 16
    eor    r9, r4      // gate 5
    orr    r4, r0      // gate 19
    eor    r1, r9, r5  // gate 7
    and    r8, r0      // gate 18
    and    r5, r2, r1  // gate 11
    bic    r4, r8      // gate 20, output C1
```

```
    orr    r5, r6      // gate 6
    eor    r2, r1      // gate 3, output S1
    eor    r5, r9      // gate 1, output C0
// r4, r5 are the output carry shares
// r7, r2 are the output sum shares
```

Listing 1.1. ARM assembly implementation of our full adder. Gate numbers in comments reference Fig. 4.

4.1 Shared Bitsliced Adder Assembly Implementation

To implement a shared bitsliced n-bit adder, the presented full adder will be repeated for n iterations. It is assumed that the shares of the slices of the operands A and B are read from memory and that the result of the sum is stored back into memory overwriting the shares of the slices of A.

In the first iteration, the secret input carry will always be 0, and therefore both input shares of C (c_{in0} and c_{in1}) will be identical. Since initializing both input carry shares to 0 would lead to a violation of the uniformity of the inputs and therefore to a leakage, c_{in0} and c_{in1} must be initialized to the same random value.

At the end of each iteration, the output (c_{out0}, c_{out1}) will become the input (c_{in0}, c_{in1}) for the next iteration. In the presented listing, this is done without requiring any additional mov instruction by simply using the same registers (r4, r5) for both (c_{out0}, c_{out1}) and (c_{in0}, c_{in1}). The use of the scratch register r0 can be optimized away by replacing it with r7 if the output s_0 is not necessary. If the output s_1 is written back to memory overwriting a_0, then no write to a_1 is required (since it will already have the correct value of s_0), further optimizing the code.

The full assembly implementation of the shared bitsliced adder is included in Appendix A.

4.2 Benchmark Results

In order to present a synthetic benchmark of the proposed implementation, the ChaCha20 cipher was implemented in a bitsliced way using the adder obtained from the NEAT algorithm and compared against the state of the art 2-shares masked implementation proposed by Jungk et al., as well as against optimized unprotected implementations. In the masked bitsliced implementation, 512 bytes (8 blocks) of keystream are generated in parallel, and therefore the best case scenarios for this implementation happen whenever the number of bytes to encrypt is a multiple of 512.

All implementations are tested on the same STM32F1 Cortex-M3 MCU, without implementing countermeasures for pipeline or MDR leakages. This way we match the protection level of related work and therefore enable a meaningful performance comparison. While this is not the focus of this paper, our implementations can be extended such that also possible non-distance-based leakage sources are eliminated. This, of course, would provide a higher security level than the mentioned related work on the expense of worse performance. Figure 5 shows the results of the benchmark both when taking into account the overhead of bitslicing inputs and outputs and when not.

Note that in situations where the number of blocks to encrypt is not a multiple of 32, bitsliced variants will suffer a degradation in performance due to their design principle. When e.g. in the worst case only one block is to be encrypted, the bitlsliced variant will be slower than conventional implementations. However, with a rising plaintext size, the performance of the bitsliced implementation increases also for non-optimal block sizes, while the throughput of non-bitlsliced alternatives remains roughly the same. This eventually leads to the bitsliced variant outperforming related work for bigger input sizes, regardless of the penalty caused by an unfavorable number of blocks. This effect is illustrated in Fig. 6.

5 Leakage Evaluation

The first-order leakage of the proposed implementation was evaluated using the t-test methodology. Two sets of traces were acquired; both using the same constant key. In the first set of traces, a constant plaintext was used. In the second set of traces, the plaintext was randomly generated with each trace. The MAPS leakage simulator was used to simulate the execution of the protected bitsliced ChaCha20 to verify that it indeed does not generate distance-based first order leakage [7]. The result plot of the MAPS leakage evaluation can be seen in Fig. 7.

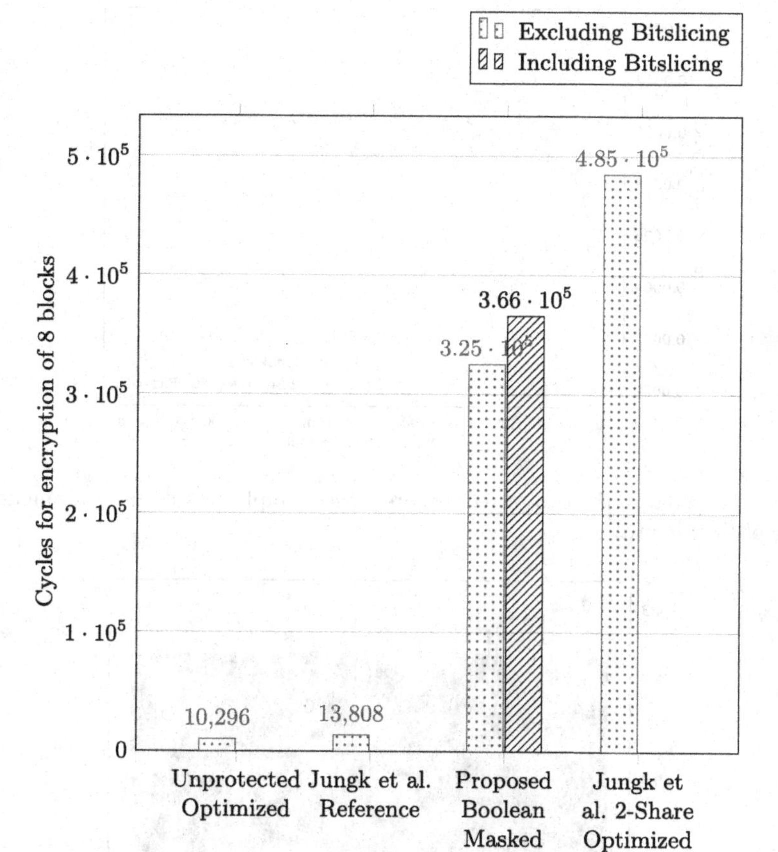

Fig. 5. Benchmark results for software implementations of the ChaCha20 algorithm using different adders

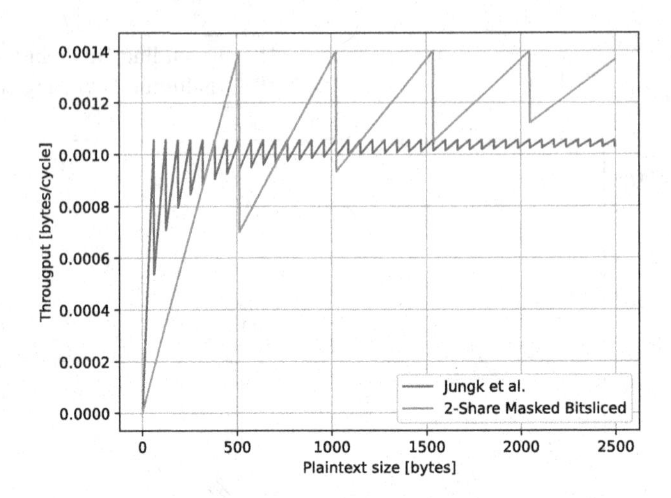

Fig. 6. Throughput of the two tested 2-share masked implementations as a function of the size of the plaintext

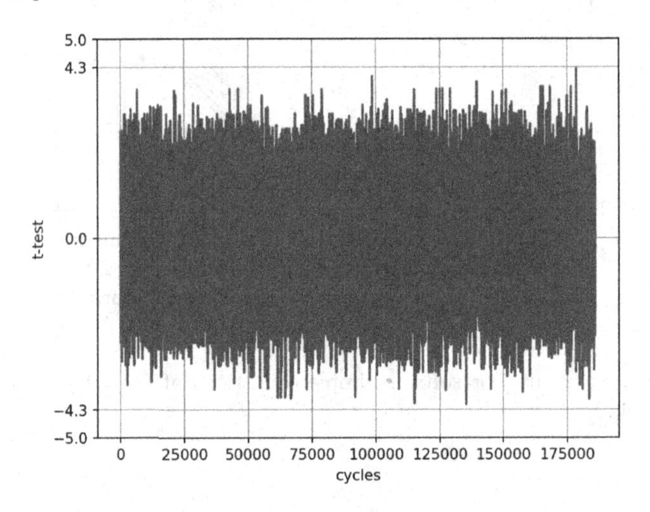

Fig. 7. Plot of the t-test performed across 1000000 power traces of the ChaCha20 encryption simulated in MAPS, with pipeline leakage simulation disabled [7]

6 Conclusion

The selection of Sparkle as one of the NIST LWC project finalists shows that ARX is a popular design pattern for LWC, especially when focusing on speed and size of a cipher. Our report highlights that protecting the modular addtion in ARX-based algorithms against basic side-channel attacks can be complicated and comes at a high cost. We presented a new neuroevolution-based approach to develop an efficient masked implementation for the modular addition. We used this optimization problem in order to evaluate the performance and usability of these evolution methods known from other research domains. Our work proves that exploiting and tinkering with neuroevolution techniques can aid in solving difficult Boolean problems. In this work, we successfully configure, extend and apply NEAT to evolve a solution for the problem of efficiently masking the modular addition. We show that the implementation generated by using neurovolution outperforms related work for larger inputs sizes.

In the future, the customized NEAT setup could be applied to similar problems. For example, we could target Hardware implementations of masked adders, especially the evolution of TI gate networks. This would require some changes, e.g. regarding the fitness and the aggregation functions. Furthermore, also other researchers could use the extended version of neat-python for tackling different Boolean problems in- or outside of the cryptography domain. Moreover, we would like to evaluate other non-neuroevolutionary but automated search algorithms. While we found standard exhaustive search methods to be too primitive to yield goods results for our adder problem setting, customized and advanced routines might be able to compete with the NEAT approach.

Acknowledgements. This project is funded by the Bavarian State Ministry of Science and the Arts and coordinated by the Bavarian Research Institute for Digital Transformation (bidt). Furthermore, this research is supported by the BayWISS Consortium Digitization.

A ARM Assembly Implementation of the Shared Bitsliced 32-bit Adder

```
.section .text
.cpu cortex-m3
.thumb
.syntax unified

.align 4
.global bitsliced_full_adder
.type bitsliced_full_adder, %function
.thumb_func
// [r0, #(4*n)]          for 0 <= n < 32   is the (n+1)-th slice of share 0 of A
// [r0, #(128 + 4*n)]    for 0 <= n < 32   is the (n+1)-th slice of share 1 of A
// [r1, #(4*n)]          for 0 <= n < 32   is the (n+1)-th slice of share 0 of B
// [r1, #(128 + 4*n)]    for 0 <= n < 32   is the (n+1)-th slice of share 1 of B
// r2 is the random refresh mask
bitsliced_full_adder:
    mov     r11, #0
    mov     r4, r2
    mov     r5, r4
_iterate_adder_next_slice:
    add     r10, r11, #0x80
    ldr     r6, [r0, r11, lsl #2]
    ldr     r7, [r1, r11, lsl #2]
    ldr     r8, [r0, r10, lsl #2]
    ldr     r10, [r1, r10, lsl #2]

// r6, r8 are shares of A
// r7, r10 are shares of B
```

```
// r4, r5 are shares of C
    eor     r9, r7, r6
    and     r6, r7
    eor     r7, r8
    eor     r8, r10
    eor     r7, r5
    eor     r10, r4
    orr     r4, r8
    eor     r12, r10, r5
    and     r7, r8
    and     r5, r9, r12
    bic     r4, r7
    orr     r5, r6
    eor     r9, r12
    eor     r5, r10

// r4, r5 are output carry shares
// r9 is the new output sum share (together with old r8)

    str     r9, [r0, r11, lsl #2]
    cmp     r11, #31
    add     r11, #1
    bne     _iterate_adder_next_slice
    bx      lr
```

References

1. Adomnicai, A., Fournier, J.J.A., Masson, L.: Bricklayer attack: a side-channel analysis on the ChaCha quarter round. In: Patra, A., Smart, N.P. (eds.) INDOCRYPT 2017. LNCS, vol. 10698, pp. 65–84. Springer, Cham (2017). https://doi.org/10.1007/978-3-319-71667-1_4
2. Adomnicai, A., Peyrin, T.: Fixslicing aes-like ciphers: New bitsliced AES speed records on arm-cortex m and RISC-V. IACR Trans. Cryptogr. Hardw. Embed. Syst. **2021**(1), 402–425 (2020). https://doi.org/10.46586/tches.v2021.i1.402-425, https://tches.iacr.org/index.php/TCHES/article/view/8739
3. Biryukov, A., Dinu, D., Le Corre, Y., Udovenko, A.: Optimal first-order Boolean masking for embedded IoT devices. In: Eisenbarth, T., Teglia, Y. (eds.) CARDIS 2017. LNCS, vol. 10728, pp. 22–41. Springer, Cham (2018). https://doi.org/10.1007/978-3-319-75208-2_2
4. Cardamone, L., Loiacono, D., Lanzi, P.L.: Evolving competitive car controllers for racing games with neuroevolution. In: Proceedings of the 11th Annual Conference on Genetic and Evolutionary Computation, pp. 1179–1186. GECCO 2009, Association for Computing Machinery, New York, NY, USA (2009). https://doi.org/10.1145/1569901.1570060
5. Coron, J.-S., Goubin, L.: On Boolean and arithmetic masking against differential power analysis. In: Koç, Ç.K., Paar, C. (eds.) CHES 2000. LNCS, vol. 1965, pp. 231–237. Springer, Heidelberg (2000). https://doi.org/10.1007/3-540-44499-8_18
6. Coron, J.-S., Großschädl, J., Tibouchi, M., Vadnala, P.K.: Conversion from arithmetic to Boolean masking with logarithmic complexity. In: Leander, G. (ed.) FSE 2015. LNCS, vol. 9054, pp. 130–149. Springer, Heidelberg (2015). https://doi.org/10.1007/978-3-662-48116-5_7
7. Le Corre, Y., Großschädl, J., Dinu, D.: Micro-architectural power simulator for leakage assessment of cryptographic software on ARM cortex-M3 processors. In: Fan, J., Gierlichs, B. (eds.) COSADE 2018. LNCS, vol. 10815, pp. 82–98. Springer, Cham (2018). https://doi.org/10.1007/978-3-319-89641-0_5

8. Deb, K., Agrawal, S., Pratap, A., Meyarivan, T.: A fast and elitist multiobjective genetic algorithm: NSGA-II. IEEE Trans. Evol. Comput. **6**(2), 182–197 (2002). https://doi.org/10.1109/4235.996017
9. Dinu, D., Großschädl, J., Corre, Y.L.: Efficient masking of ARX-based block ciphers using carry-save addition on Boolean shares. In: Nguyen, P.Q., Zhou, J. (eds.) Information Security - 20th International Conference, ISC 2017, Ho Chi Minh City, Vietnam, November 22–24, 2017, Proceedings. LNCS, vol. 10599, pp. 39–57. Springer, Cham (2017). https://doi.org/10.1007/978-3-319-69659-1_3
10. Floreano, D., Dürr, P., Mattiussi, C.: Neuroevolution: from architectures to learning. Evol. Intel. **1**(1), 47–62 (2008)
11. Goubin, L.: A sound method for switching between Boolean and arithmetic masking. In: Koç, Ç.K., Naccache, D., Paar, C. (eds.) Cryptographic Hardware and Embedded Systems - CHES 2001, Third International Workshop, Paris, France, May 14–16, 2001, Proceedings. LNCS, vol. 2162, pp. 3–15. Springer, Cham (2001). https://doi.org/10.1007/3-540-44709-1_2
12. Groß, H., Stoffelen, K., Meyer, L.D., Krenn, M., Mangard, S.: First-order masking with only two random bits. In: Bilgin, B., Petkova-Nikova, S., Rijmen, V. (eds.) Proceedings of ACM Workshop on Theory of Implementation Security Workshop, TIS@CCS 2019, London, UK, 11 November 2019, pp. 10–23. ACM (2019). https://doi.org/10.1145/3338467.3358950
13. Hausknecht, M., Lehman, J., Miikkulainen, R., Stone, P.: A neuroevolution approach to general Atari game playing. IEEE Trans. Comput. Intell. AI Games **6**(4), 355–366 (2014). https://doi.org/10.1109/TCIAIG.2013.2294713
14. Jungk, B., Petri, R., Stöttinger, M.: Efficient side-channel protections of ARX ciphers. IACR Trans. Cryptogr. Hardw. Embed. Syst. **2018**(3), 627–653 (2018). https://doi.org/10.13154/tches.v2018.i3.627-653, https://tches.iacr.org/index.php/TCHES/article/view/7289
15. McIntyre, A., Kallada, M., Miguel, C.G., da Silva, C.F.: Neat-python. https://github.com/CodeReclaimers/neat-python
16. Nadkarni, J., Ferreira Neves, R.: Combining neuroevolution and principal component analysis to trade in the financial markets. Expert Syst. App. **103**, 184–195 (2018). https://doi.org/10.1016/j.eswa.2018.03.012, https://www.sciencedirect.com/science/article/pii/S0957417418301519
17. Risi, S., Hughes, C.E., Stanley, K.O.: Evolving plastic neural networks with novelty search. Adapt. Behav. **18**(6), 470–491 (2010). https://doi.org/10.1177/1059712310379923
18. Schneider, T., Moradi, A., Güneysu, T.: Arithmetic addition over Boolean masking. In: Malkin, T., Kolesnikov, V., Lewko, A.B., Polychronakis, M. (eds.) ACNS 2015. LNCS, vol. 9092, pp. 559–578. Springer, Cham (2015). https://doi.org/10.1007/978-3-319-28166-7_27
19. Schwabe, P., Stoffelen, K.: All the AES you need on cortex-M3 and M4. In: Avanzi, R., Heys, H. (eds.) SAC 2016. LNCS, vol. 10532, pp. 180–194. Springer, Cham (2017). https://doi.org/10.1007/978-3-319-69453-5_10
20. Stanley, K.O., Miikkulainen, R.: Evolving neural network through augmenting topologies. Evol. Comput. **10**(2), 99–127 (2002). https://doi.org/10.1162/106365602320169811
21. Stanley, K.O., Miikkulainen, R.: Competitive coevolution through evolutionary complexification. J. Artif. Intell. Res. **21**, 63–100 (2004)

22. Stanley, K., Bryant, B., Miikkulainen, R.: Real-time neuroevolution in the Nero video game. IEEE Trans. Evol. Comput. **9**(6), 653–668 (2005). https://doi.org/10.1109/TEVC.2005.856210
23. Yao, X.: Evolving artificial neural networks. Proc. IEEE **87**(9), 1423–1447 (1999). https://doi.org/10.1109/5.784219
24. Zadeh, L.: Optimality and non-scalar-valued performance criteria. IEEE Trans. Autom. Control **8**(1), 59–60 (1963). https://doi.org/10.1109/TAC.1963.1105511

Explainable AI and Deep Autoencoders Based Security Framework for IoT Network Attack Certainty (Extended Abstract)

Chathuranga Sampath Kalutharage(✉), Xiaodong Liu, and Christos Chrysoulas

Edinburgh Napier University, Scotland, UK
{c.kalutharage,x.liu,c.chrysoulas}@napier.ac.uk
https://www.napier.ac.uk/

Abstract. Over the past few decades, Machine Learning (ML)-based intrusion detection systems (IDS) have become increasingly popular and continue to show remarkable performance in detecting attacks. However, the lack of transparency in their decision-making process and the scarcity of attack data for training purposes pose a major challenge for the development of ML-based IDS systems for Internet of Things (IoT). Therefore, employing anomaly detection methods and interpreting predicted results in terms of feature contribution or performing feature-based impact analysis can increase stakeholders confidence. To this end, this paper presents a novel framework for IoT security monitoring, combining deep autoencoder models with Explainable Artificial Intelligence (XAI), to verify the credibility and certainty of attack detection by ML-based IDSs. Our proposed approach reduces the number of black boxes in the ML decision-making process in IoT security monitoring by explaining why a prediction is made, providing quantifiable data on which features influence the prediction and to what extent, which are generated from SHaply Adaptive values exPlanations (SHAP) linking optimal credit allocation to local explanations. This was tested using the USB-IDS benchmark dataset and a detection accuracy of 84% (benign) and 100% (attack) was achieved. Our experimental results show that integrating XAI with the autoencoder model obviates the need of malicious data for training purposes, but can provide attack certainty for detected anomalies, proving the validity of the proposed methodology.

Keywords: IoT Security · Anomaly detection · Explainable AI

1 Introduction

Since IoT devices are connected through the Internet, there is a high possibility that they are vulnerable to cyberattacks such as impersonate, interception and penetration by unauthorized users and viruses [24]. So these devices require a proper security mechanism. Since traditional signature-based intrusion detection

© The Author(s), under exclusive license to Springer Nature Switzerland AG 2022
W. Li et al. (Eds.): ADIoT 2022, LNCS 13745, pp. 41–50, 2022.
https://doi.org/10.1007/978-3-031-21311-3_8

systems (IDS) are no longer effective at detecting attacks, as modern attacks are sophisticated and complex, most IoT security research is currently based on Artificial Intelligence. Since ML systems are iterative and dynamic, advanced solutions based on ML are better suited to detect and mitigate the impact of cyberattacks and potential threats to IoT data and infrastructures [13].

Most of the ML-based solutions proposed in the literature are supervised learning methods that require labeled training data on attack and benign activities with certainty in ground truth. However, labeled attack data is expensive to obtain, and legal, ethical, and privacy concerns may not allow realistic data to be shared across research communities. Therefore, the use of anomaly-based detection methods is encouraged in the security field, as these models can be trained using benign data only. The main drawback of the anomaly-based method is that it often triggers false positives since it flags all unusual patterns as potential attacks even when they are not [9]. Understanding the reasons for instance prediction can reduce these false alarms and be the first step for domain experts to make decisions to prevent future attacks. Moreover, most ML-based mechanisms in security applications solve the attack detection problem and only give results whether it is an attack (anomaly) or not, and often work as a black box for the end user without providing much details on their decision-making process [20]. As a result, in operational environments, interpreting IDS outputs from the operator's point of view and transferring them into actionable reports is a challenge. Therefore, explaining the reasons behind a model's decisions has become an integral part of IDS solutions as ML becomes much more widely used in the operation of critical systems, to the point that governments are beginning to include it in legislation [11]. The ML community has recently concentrated on developing XAI methods that are easier for users to understand [12]. XAI uses natural language explanations and visualizations to show how the machine learning model arrived at its decisions.

To overcome the aforementioned limitations in IoT security monitoring, this paper presents a novel framework, combining deep autoencoder models with XAI, to identify the most influential features of anomalous behavior that violates predefined cybersecurity policies. Explainable models help to understand and diagnose the decisions made by the model, thereby increasing confidence in the data-driven IoT network security model. A domain expert can easily interpret the decisions offered by explainable models since it simplifying the knowledge discovery process. The main contributions of this research therefore are as follows. The model will detect anomalies in the IoT network and the model will demonstrate the certainty of the detected anomaly rather than providing a false attack. Additionally, the model will demonstrate the most influential features with a weight for each anomalous behavior. This model decision-making process (model explainability) can be mapped to domain expert knowledge for greater attack certainty. Thus, consequently, the model meets all the fundamental needs of modern IoT networks, providing accurate, reliable and transparent anomaly detection.

The rest of the paper is organized as follows: Section 2 presents an overview of background and related work. Section 3 describes the proposed Explainable AI and Deep Autoencoders based methodology. Section 4 describes the experimental results carried out using the USB-IDS benchmark data set and finally Sect. 5 concludes the paper.

2 Background and Related Work

This research focuses on the development of a security framework for IoT security monitoring, combining deep autoencoders with XAI. Therefore, the work associated with each area is discussed separately in this section.

2.1 Explainable AI (XAI)

An Explainable AI (XAI) system aims to make its behavior more understandable to humans by providing explanations. There are several XAI concepts that can be used to help develop more efficient and human-understandable AI systems [3]. The XAI system should be able to describe its capabilities and concepts, as well as what it has done, what it is currently doing, and what will happen next. It must also be able to reveal the key information on which it acts on [3]. Several ML-based IDSs have been proposed over the past decades to protect cyber networks from malicious threat actors with exceptional performance [23]. However, these complex models are often known as black box models and difficult to understand for end users. In the context of security, a single incorrect IDS prediction can expose systems and networks to major cyber risks. Therefore, XAI should be integrated with traditional IDS to enhance its credibility and reliability. Mahbooba et al. presented on explaining each predicted outcome by extracting rules from the decision tree trained and tested on the dataset. Only the expected results and the overall model response were explained using these extracted rules [19]. Similarly, Sinclair et al. and Ojugo et al. presented two separate papers to improve model performance [21,26]. In this work, rules were derived using decision trees and genetic algorithms (GA). Instead of having an optimal rule, authors argued that IDSs should be created using a set of rules generated by machine learning. This concept was further expanded by Dais et al. by making decision-making processes more transparent [8]. However, none of these works focus on improving the IDS using the explanations of XAI tools.

2.2 Unsupervised Model Explanations

Clustering is a popular technique for solving unsupervised learning problems. The issue of cluster interpretability has had a poor track record of success [5]. A widely used explanation is to represent a cluster of points by their centroid or by a group of distant points in the collection [22]. When the clusters are compact or isotropic it works fine, but it fails in all other cases. Due to complex patterns in data distributions, it is unrealistic to expect isotropic data in the cyber domain. Another popular technique is to use principal component analysis (PCA) projections or T-distributed Stochastic Neighbourhood Embedding (t-SNE) to visualise clusters in a two-dimensional network [18]. But, the connection between the clusters and the original variables is obscured by the reduction in the dimensionality of the features. Van der Maaten et al. suggest Interpretable Clustering via Optimal Trees (ICOT), in which the clusters are represented by the leaves, and decision tree (unsupervised) built using feature values [4]. Liu et al. and Lundberg et al. presented two different papers on clustering method based on decision trees [15,17]. Both papers present a method that builds explainable clusters instead of explaining clusters generated by algorithms. Due to the aforementioned limitations, clustering would not be a suitable unsupervised method for our problem.

2.3 Explaining Anomalies

In the field of cybersecurity, unsupervised learning techniques such as anomaly detection are gaining popularity because a large number of labeled attack examples are needed for supervised learning, and new types of attacks will continue to emerge [25]. Almost as important as the model's predictive accuracy is the capacity to explain an anomaly detection methodology in critical sectors, such as infrastructure security [1]. Therefore, an effective anomaly explanation will greatly increase the usefulness of anomaly detection methods in real-world applications. Explaining outliers can significantly reduce the need of manual inspection of false alarms by security analysts. Goodal et al. presented a system for detecting and interpreting streaming anomalies in computer network traffic and logs, visualization of the contexts of the anomaly serves as the basis for the explanation [10]. Liu et al. presented a new Contextual Outlier Interpretation (COIN) method to explain existing outlier anomalies spotted by detectors [16]. Collaris created two dashboards using a combination of state-of-the-art explanatory techniques. These two dashboards allow the domain expert to understand the prediction. Explanations are based on currently used explanation techniques, including partial dependency diagrams, instance-level feature importance method, and local rule mining (a variant of LIME). Other research presents an SVM-based malware detection and explanation approach to explaining output made by recognizing the features that most strongly influence detection and verifying if the extracted features that influence a detection match common vulnerable characteristics [2]. Valerio La Gatta et al. presented the local explanation method CASTLE (Cluster-Aided Space Transformation for Local explanations), which provides decision rules proposing how the model prediction can be generalized to unseen instances and provides local information about the importance of the feature [14]. However, none of the above studies explained the IoT network anomalies detected by autoencoders, therefore, our work is unique and different from the above studies.

3 Methodology

To the best of our knowledge, most explainable approaches are developed for supervised learning methods (classification algorithms). But unlike existing approaches to explain a prediction, our goal is to develop an approach to explain an anomaly detected by an autoencoder model in the context of IoT security monitoring. To this end, we use the reconstruction error (see Eq. 1) of an autoencoder model to define IoT network anomalies (Anomaly Score). Anomalies are instances with a high reconstruction error values. In other words, a high difference (error) between the input and output (reconstructed) value is known as an anomaly. A threshold for the reconstruction error is estimated using a benign training dataset. If an anomaly exists in the incoming data, the explanatory model should be able to explain why this instance could not be well predicted (reconstructed) by the autoencoder model. As a result, the error is linked to an explanation and the proposed method calculates the SHAP values of the output features and compares them to the true (anomalous) values of the input.

$$L(A, A') = \sum_{i=1}^{n} (a_i - a_i')^2 \tag{1}$$

Equation 1 denotes the computation of reconstruction error in our work. Given input row A with a set of features a_i, and its output A'' with reconstructed feature values a_i', and using an autoencoder model f, the reconstruction error of row is sum of the reconstruction errors of each feature. Then the features in error list need to be reordered in a descending order such that $|a_1 - a_1'| > |a_n - a_n'|$, to find top R features which includes a set of selected features for which the total corresponding errors indicate a modifiable percentage of $L(A, A')$. The model uses SHAP[1] values to describe which features were responsible for each of the high reconstruction errors in top R features.

In the explanation process, we first detect the anomalous instance using the model. Then we take the features with the highest reconstruction error and save them in the top R feature list. To get the SHAP values of each feature (i.e. a_i) in the list, Kernel SHAP is used. Then the result is displayed in a two-dimensional array, in which each of the rows represents the SHAP values for features in the top R features. The model divides SHAP values into two categories in the next step. One of the categories corresponds to contributing values that push the predicted value away from the input value and the other category is offsetting values that push the predicted value towards the true value. The division process is as follows. If the value of the input feature is greater than the output value, negative SHAP values are contributing features and positive values are offset features. If the output feature value is greater than the input value, positive SHAP values are contributing features and negative shape values are offset features. These steps return two list those are SHAP contributing and SHAP offsetting.

[1] https://www.kaggle.com/code/dansbecker/shap-values.

Finally, it selects the features with high SHAP values of the features in the top R features. From each row of contributing and offsetting SHAPs, we extract the highest values. Our goal is to explain the result with the most influencing features to the user to understand the reason for the anomaly. Figure 1 illustrates the proposed approach.

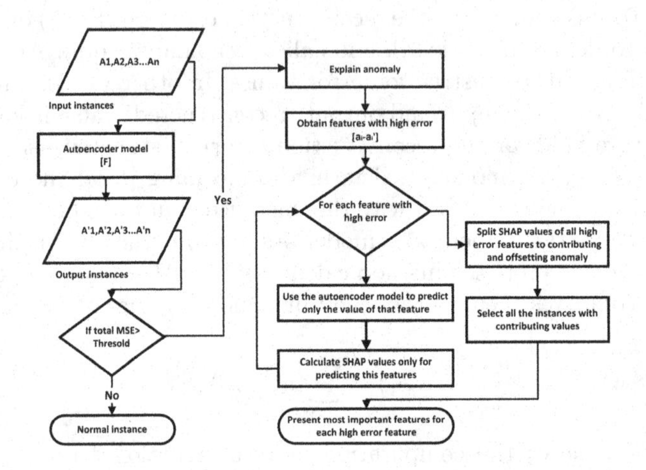

Fig. 1. Process of explanation of each anomaly detected by an autoencoder

4 Experimental Evaluation

4.1 Dataset

The USBIDS dataset [6] was used in our experimental evaluation because it provides clear feature descriptions compared to other alternative datasets. It consist of 17 csv files of labelled network flow data. A combination of denial of service (DoS) attack and defensive module consists of 16 files in addition to a benign (unaffected by an attack) network traffic data file. CIC FlowMeter[2] has used to derive network flows in the dataset. The naming convention of the 16 non-normative CSV files helps to identify the collection scenario. For example, TCPFlood-NoDefense.csv provides flows obtained by executing TCPFlood with no defense in place.

4.2 Experimental Setting

We trained the model using benign data only and two sets of attack data together with benign data were used to evaluate the model. A fully connected autoencoder model with RELU activation was used. To lighten the model, only 2 hidden

[2] https://github.com/ahlashkari/CICFlowMeter.

layers are used in the network. The hidden layers contain 10 and 32 neurons respectively. Using benign data, the maximum Mean squared error(MSE) value set as the anomaly threshold. The proposed algorithm was implemented using Python 3.8 with TensorFlow and the Keras library. 40 epochs and a learning rate of 0.01 were used with the Adam optimizer. The experiments were run on a ZenBook 2.30 GHz Intel Core i7 with 16 GB of RAM.

Table 1. Results on Hulk Attack of USBIDS dataset with a comparison (Recall) to the current state of the art [7]

Detection method	Attack Hulk no defence	Attack Hulk Evasive	Attack Hulk reqtimeout
DT [7]	0.97	0.06	0.97
RF [7]	0.98	0.00	0.98
DNN [7]	0.67	0.05	0.66
Proposed method	0.98	1.0	1.0

4.3 Results and Discussion

We experimented with different models to find the best performing model with the lightest architecture. Among them, the above model performed the best and the results are shown in Table 1 with a comparison to the current state of the art. In recent years, many tools and libraries are released to open black box models. However, there are no standard performance metrics to compare the performance of such algorithms. No single explainability method is better than the others. Thus, to evaluate the proposed model, we mapped XAI outcomes of our model to domain expertise. To this end, we consulted three cybersecurity experts and presented the feature set of the dataset together with attack types and asked them to rank the importance/influence of each feature in detecting the attack. For example, according to the domain experts, Forward packets per second (Fwd Packets/s), Backward packets per second (Bwd Packets/s), Flow packets per second (Flow Packets/s), Backward Packet Length Max (Bwd Packet Length Max), Packet Length Max are the most influential features in detecting a DoS attack, which comply with the out of the proposed approach. Further to model evaluation, such a list can be used in our approach to compare the XAI output with that list to further verify the certainty of the detected anomaly as an attack.

After deploying our model, we get anomalous instances as output. These anomalous instances are explained by the explainable model with an influence weight as shown in Fig. 3. According to this explanation, forward packets per second (Fwd packets/s) is the most influential feature (contribute) with a weight value of 0.0882. After that, Backward Packets per second, Flow Packets per second, Backward Packet Length Max have an effect with their respective values 0.0845, 0.0749, 0.016. Forward Packet Length Standard (Fwd Packet Length Std) is the offsetting feature for this anomalous instance. This offsetting features do not affect to the attack certainty as they are not contributing to the mean squared error.

We found forwarding packets per second (FWD packets/s) as one of the most influencing features by using our explainable model (anomaly instance 548271) and expertise knowledge. Then we compared the feature value of forward packets per second (FWD packets/s) respectively benign and attack classes. Packets per second feature has a value ranging between 0 and 3000 for a benign class, but in the attack state, this feature value increases up to 8000 per second. Backward packet per second also showing similar result, backward packets per second feature values vary from 0 to 3500 benign states and up to 8000 packets per second in the attack state (Fig. 2).

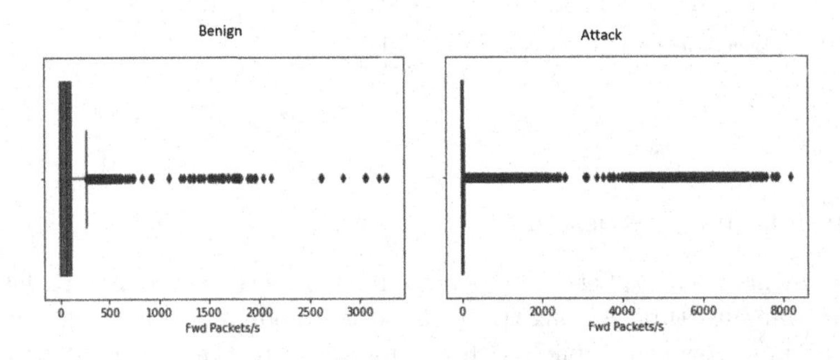

Fig. 2. Packets per second feature values of benign and attack classes

Considering these facts, we can confirm that most of the influenced features explained by our model are correct. Finally, considering these results, we can confirm that our explainable model is more efficient in finding the attack certainty of the anomaly that is detected by the existing anomaly detection method.

5 Conclusion

ML-based IDSs are attracting a lot of attention from security researchers, but have limited use in the operational environment due to their black box nature. It is unclear what things contribute to their decisions, and most anomaly detection detects the anomalies, but there is no certainty about the attack. To address these issues, we have proposed a framework in which instance-wise explanations, local and global explanations, and relationships between features and system outcomes help in obtaining key decision-making features, which will eventually lead to estimate the attack certainity. By analysing the model explanations, the cybersecurity expert will also be able to make the final decision regarding the anomaly. In addition, the explanations allow the end user to better understand

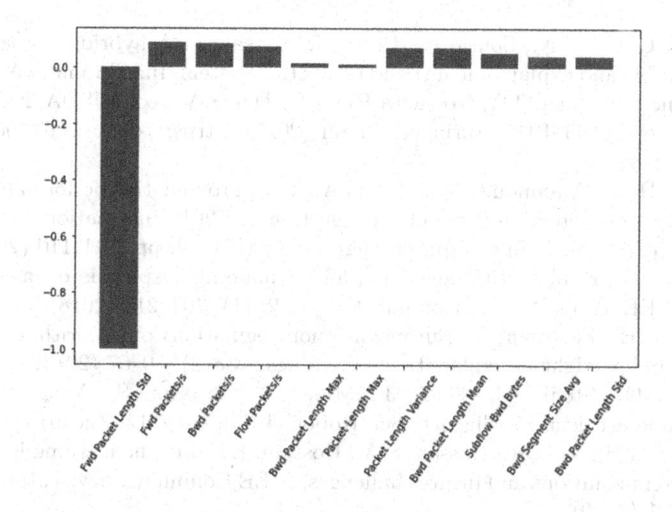

Fig. 3. Feature influenced of the selected anomalous anomalous instance (548271)

the decision and influencing features with weighting. In the future, we plan to extend this work to map XAI outputs to local security policies in the IoT network to detect which policy is being violated by the reported anomaly. In operational environments, this will certainly be useful for interpreting IDS outputs from the operator's point of view and transferring them into actionable reports. Additionally, this framework will be deployed in a real IoT network environment to investigate its capabilities in production environments.

References

1. Amarasinghe, K., Kenney, K., Manic, M.: Toward explainable deep neural network based anomaly detection. In: 2018 11th International Conference on Human System Interaction (HSI), pp. 311–317. IEEE (2018)
2. Arp, D., Spreitzenbarth, M., Hubner, M., Gascon, H., Drebin, K.: Effective and explainable detection of android malware in your pocket. In: Network and Distributed System Security Symposium, pp. 1–15 (2014)
3. Bellotti, V., Edwards, K.: Intelligibility and accountability: human considerations in context-aware systems. Hum. Comput. Interact. **16**(2–4), 193–212 (2001)
4. Bertsimas, D., Dunn, J.: Optimal classification trees. Mach. Learn. **106**(7), 1039–1082 (2017). https://doi.org/10.1007/s10994-017-5633-9
5. Bertsimas, D., Orfanoudaki, A., Wiberg, H.: Interpretable clustering via optimal trees. arXiv preprint arXiv:1812.00539 (2018)
6. Catillo, M., Del Vecchio, A., Ocone, L., Pecchia, A., Villano, U.: USB-IDS-1: a public multilayer dataset of labeled network flows for IDS evaluation. In: 51st Annual IEEE/IFIP International Conference on Dependable Systems and Networks Workshops (DSN-W), pp. 1–6. IEEE (2021)
7. Catillo, M., Del Vecchio, A., Pecchia, A., Villano, U.: Transferability of machine learning models learned from public intrusion detection datasets: the cicids2017 case study. Software Qual. J. 1–27 (2022)

8. Dias, T., Oliveira, N., Sousa, N., Praça, I., Sousa, O.: A hybrid approach for an interpretable and explainable intrusion detection system. In: Abraham, A., Gandhi, N., Hanne, T., Hong, TP., Nogueira Rios, T., Ding, W. (eds.) ISDA 2021. LNNS, vol. 418, pp. 1035–1045. Springer, Cham (2022). https://doi.org/10.1007/978-3-030-96308-8_96
9. Elshafie, H.M., Mahmoud, T.M., Ali, A.A.: Improving the performance of the snort intrusion detection using clonal selection. In: 2019 International Conference on Innovative Trends in Computer Engineering (ITCE), pp. 104–110 (2019)
10. Goodall, J.R., et al.: Situ: identifying and explaining suspicious behavior in networks. IEEE Trans. Visual. Comput. Graph. **25**(1), 204–214 (2018)
11. Goodman, B., Flaxman, S.: European union regulations on algorithmic decision-making and a "right to explanation". AI Magaz. **38**(3), 50–57 (2017)
12. Gunning, D., Stefik, M., Choi, J., Miller, T., Stumpf, S., Yang, G.Z.: Xai-explainable artificial intelligence. Sci. Robot. **4**(37), eaay7120 (2019)
13. Hussain, F., Hussain, R., Hassan, S.A., Hossain, E.: Machine learning in IoT security: current solutions and future challenges. IEEE Commun. Surv. Tutorials **22**(3), 1686–1721 (2020)
14. La Gatta, V., Moscato, V., Postiglione, M., Sperli, G.: Castle: cluster-aided space transformation for local explanations. Expert Syst. Appl. **179**, 115045 (2021)
15. Liu, B., Xia, Y., Yu, P.S.: Clustering through decision tree construction. In: Proceedings of the ninth international conference on Information and knowledge management, pp. 20–29 (2000)
16. Liu, N., Shin, D., Hu, X.: Contextual outlier interpretation. arXiv preprint arXiv:1711.10589 (2017)
17. Lundberg, S.M., Erion, G.G., Lee, S.I.: Consistent individualized feature attribution for tree ensembles. arXiv preprint arXiv:1802.03888 (2018)
18. Van der Maaten, L., Hinton, G.: Visualizing data using T-SNE. J. Mach. Learn. Res. **9**(11) (2008)
19. Mahbooba, B., Timilsina, M., Sahal, R., Serrano, M.: Explainable artificial intelligence (xai) to enhance trust management in intrusion detection systems using decision tree model. Complexity 2021 (2021)
20. Marino, D.L., Wickramasinghe, C.S., Manic, M.: An adversarial approach for explainable AI in intrusion detection systems. In: IECON 2018–44th Annual Conference of the IEEE Industrial Electronics Society, pp. 3237–3243. IEEE (2018)
21. Ojugo, A., Eboka, A., Okonta, O., Yoro, R., Aghware, F.: Genetic algorithm rule-based intrusion detection system (gaids). J. Emerg. Trends Comput. Inform. Sci. **3**(8), 1182–1194 (2012)
22. Radev, D.R., Jing, H., Styś, M., Tam, D.: Centroid-based summarization of multiple documents. Inform. Process. Manage. **40**(6), 919–938 (2004)
23. Salih, A.A., Abdulazeez, A.M.: Evaluation of classification algorithms for intrusion detection system: a review. J. Soft Comput. Data Mining **2**(1), 31–40 (2021)
24. Samaila, M.G., Neto, M., Fernandes, D.A., Freire, M.M., Inácio, P.R.: Challenges of securing internet of things devices: a survey. Secur. Privacy **1**(2), e20 (2018)
25. Siddiqui, M.A., et al.: Detecting cyber attacks using anomaly detection with explanations and expert feedback. In: ICASSP 2019-2019 IEEE International Conference on Acoustics, Speech and Signal Processing (ICASSP), pp. 2872–2876 (2019)
26. Sinclair, C., Pierce, L., Matzner, S.: An application of machine learning to network intrusion detection. In: Proceedings 15th Annual Computer Security Applications Conference (ACSAC 1999), pp. 371–377 (1999)

Constraints and Evaluations on Signature Transmission Interval for Aggregate Signatures with Interactive Tracing Functionality

Ryu Ishii[1](✉), Kyosuke Yamashita[2,3], Zihao Song[4], Yusuke Sakai[2], Tadanori Teruya[2], Goichiro Hanaoka[2], Kanta Matsuura[1], and Tsutomu Matsumoto[2,4]

[1] The University of Tokyo, Tokyo, Japan
ryuishii@iis.u-tokyo.ac.jp
[2] National Institute of Advanced Industrial Science and Technology, Tokyo, Japan
[3] Osaka University, Suita, Japan
[4] Yokohama National University, Yokohama, Japan

Abstract. Fault-tolerant aggregate signature (FT-AS) is a special type of aggregate signature that is equipped with the functionality for tracing signers who generated invalid signatures in the case an aggregate signature is detected as invalid. In existing FT-AS schemes (whose tracing functionality requires multi-rounds), a verifier needs to send a feedback to an aggregator for efficiently tracing the invalid signer(s). However, in practice, if this feedback is not responded to the aggregator in a sufficiently fast and timely manner, the tracing process will fail. Therefore, it is important to estimate whether this feedback can be responded and received in time on a real system.

In this work, we measure the total processing time required for the feedback by implementing an existing FT-AS scheme, and evaluate whether the scheme works without problems in real systems. Our experimental results show that the time required for the feedback is 605.3 ms for a typical parameter setting, which indicates that if the acceptable feedback time is significantly larger than a few hundred ms, the existing FT-AS scheme would effectively work in such systems. However, there are situations where such feedback time is not acceptable, in which case the existing FT-AS scheme cannot be used. Therefore, we further propose a novel FT-AS scheme that does not require any feedback. We also implement our new scheme and show that a feedback in this scheme is completely eliminated but the size of its aggregate signature (affecting the communication cost from the aggregator to the verifier) is 144.9 times larger than that of the existing FT-AS scheme (with feedbacks) for a typical parameter setting, and thus has a trade-off between the feedback waiting time and the communication cost from the verifier to the aggregator with the existing FT-AS scheme.

Keywords: Sensor networks · Aggregate signature

© The Author(s), under exclusive license to Springer Nature Switzerland AG 2022
W. Li et al. (Eds.): ADIoT 2022, LNCS 13745, pp. 51–71, 2022.
https://doi.org/10.1007/978-3-031-21311-3_3

1 Introduction

An aggregate signature scheme [1,3,8,19] is a signature scheme in which multiple signatures can be aggregated into a single aggregate signature. Aggregate signature schemes potentially allow us to significantly reduce the communication cost in information systems where a large number of signed data is transmitted. For instance, an aggregate signature scheme can be used for alive monitoring of devices in a factory. Suppose that there are thousands of devices that send their own signatures as survival reports. Thanks to an aggregate signature scheme, the thousands of reports can be aggregated into a single aggregate signature. Thus, the factory can pass the entire reports to the monitoring center by sending the aggregate signature, which will drastically reduce the bandwidth consumption.

Here, however, a potential problem is that, if invalid signatures are contained in the aggregate signature, then the entire reports would be rejected. In this paper, we call signers who generate invalid signatures *rogue* signers. Rogue signers can be seen to model signers who generate invalid signature probabilistically (due to, e.g., the failure of devices), or signers that are corrupted by a malicious party.[1] At first glance, it seems trivial to detect invalid signatures if an aggregator verifies individual signatures before aggregation. In practice, however, aggregators are assumed to be relatively small devices on the communication path, such as IoT devices, which have small memory size and would be difficult to verify many signatures sufficiently fast. Therefore, we assume aggregators are inexpensive small devices that can aggregate fast but cannot verify individual signatures sufficiently fast before aggregation or store them all.

For resolving this problem, Hartung et al. [10] proposed Fault-Tolerant Aggregate Signature (FT-AS for short) scheme which is equipped with the functionality for identifying rogue signers. In FT-AS schemes, the aggregator temporarily stores all signatures which have not been aggregated in its memory, and when an invalid aggregate signature is detected, it identifies the rogue signer(s) by using combinatorial methods such as cover-free family [13,15,30], or group testing [5,6]. Note that these schemes are more efficient than the trivial scheme because the combinatorial methods require light computation and less times of verification than trivial method. A potential problem in these schemes is that the aggregator is required to have a relatively large memory which can store all individual (non-aggregated) signatures. Since the aggregator is assumed to be a non-powerful device, it is not always the case that such a sufficiently large memory is available to the aggregator. Therefore, following [10], there have seen some attempts to reduce the required memory size for the aggregator in the context of FT-AS schemes. Especially, [12] investigated an approach for reducing the aggregator's memory size by introducing a *multi-round tracing* [26]. This special type of FT-AS was called *aggregate signature with interactive tracing functionality (ASIT)* in [12]. As mentioned above, one possible application of FT-AS is alive monitoring of edge devices in a factory, where each edge device periodically

[1] In this paper, we mainly focus on signers that probabilistically generate invalid signatures. However, we can think of them as corrupted by a malicious party.

sends a signature and a detecting process for the rogue signers is carried out by using multiple aggregated signatures at each time period. In such a scenario, in contrast to the single-round setting, FT-AS schemes in the multi-round setting do not identify rogue signers in a single execution of the tracing functionality. Typically, it works as follows: Once the aggregator receives individual signatures, it aggregates them by following some predetermined way (by the underlying combinatorial method such as Dynamic Traitor Tracing (DTT) [7]), and then sends the aggregate signatures to the verifier. The verifier, on receiving the aggregate signatures, verifies them to determine the next way of aggregation (by running the combinatorial method), and sends it to the aggregator as a *feedback*. When the aggregator receives individual signatures in the next time period, it aggregates them based on the way that is determined by the feedback. This process is repeated until all rogue signers are identified.

As mentioned above, in existing multi-round FT-AS schemes, the verifier sends a feedback to the aggregator for efficiently tracing rogue signers. Thus, in real systems, this feedback must be transmitted in a sufficiently fast and timely manner to the aggregator so that the aggregator can proceed to the next step of the tracing procedure in time. Therefore, it is important to investigate whether this feedback can be sent and received sufficiently fast on a real system, and might be necessary to consider an alternative scheme if the feedback waiting time is not acceptable.

Our Contribution. To the best of our knowledge, the most efficient multi-round FT-AS scheme is the one proposed by Ishii et al. [12]. (In the following, we refer to this scheme as AS-FT-2, following the description in [12].) The first contribution of our work is that we measure the total processing time required for the feedback by implementing AS-FT-2. In order to realize an environment in cyberspace that is as close to a real system as possible, we implement this scheme in a simulation environment built on Amazon Web Services (AWS). We measure the performance of the algorithms, and find out that the feedback waiting time is 605.3 ms on average in the setting with the number of signers $N = 3000$ including $d = 5$ rogue signers (for which we have the above mentioned example of a factory in mind). The feedback waiting time is asymptotically proportional to N. This indicates that in applications whose acceptable feedback time is significantly larger than a few hundred ms, the existing FT-AS scheme can be used without problems. However, these schemes cannot be used if applications require faster feedback time. For instance, the average requirement for the transmission interval is considered to be 5 seconds to an hour [28], whereas in some real-time communication systems, the data transmission interval for cooperative intelligent transportation systems is considered to be 200 ms [29]. The existing scheme would not be suitable for the latter case.

Our second contribution is that we propose a new variant of an ASIT scheme [12] that does not require a feedback, in anticipation of applications where the feedback waiting time is shorter so that existing schemes cannot be used. The idea behind our proposed scheme is that we let the aggregator decide how to generate aggregate signatures beforehand. More specifically, our scheme employs Sequential Traitor Tracing (STT) [23] as a tracing functionality, instead of DTT employed in the existing ASIT scheme. (See Sect. 3 for the technical details.) We also implement our new scheme, and compare the performances with AS-FT-2. The experiment reveals that the new scheme runs faster than the existing scheme, but requires more communication cost (i.e., the bandwidth for sending aggregate signatures). See Sect. 5 for the discussion.

Related Work. Boneh et al. [3] proposed the first aggregate signature scheme (together with its concept itself), which is secure in the random oracle model and based on the BLS signature [4] in groups with efficiently computable bilinear maps. Hohenberger et al. [11] gave an aggregate signature scheme using multilinear maps in the standard model. These schemes can aggregate individual signatures as well as already aggregated signatures in any order.

Hartung et al. [10] proposed fault-tolerant aggregate signature schemes. In their schemes, a cover-free family [13,15,30], which is a combinatorial scheme, is used to determine sets of individual signatures to be aggregated. Several works [24–26] provide efficient aggregate signatures with a tracing functionality based on group testing [5,6]. These schemes are shown to be secure against static attackers.

There are other types of aggregate signature schemes. One is sequential aggregate signature, which is first proposed by Lysyanskaya et al. [19] and secure in the random oracle model. Since then, a number of schemes have been proposed both in the random oracle model [9,19,21] and in the standard model [17,18]. Another type is aggregate signature with synchronized aggregation, which is first proposed by Gentry and Ramzan [8] (in the identity-based setting) and secure in the random oracle model. Again, since then, several constructions have been proposed both in the random oracle model [1,8] and in the standard model [1].

Paper Organization. In Sect. 2, we introduce notations and recall the definitions of ASIT. In Sect. 3, we review the ASIT scheme AS-FT-2 by Ishii et al. [12], observe the feedback waiting time of this scheme, and give a discussion on it. In Sect. 4, we propose an ASIT scheme, called AS-SW-1, that does not require a feedback, based on a STT. In Sect. 5, we show the experimental results for the proposed scheme and discuss which of AS-FT-2 and AS-SW-1 is more suitable for some network systems. Finally, Sect. 6 concludes this work.

2 Preliminaries

2.1 Notations

We let $[n] := \{1, \ldots, n\}$. We denote the empty string by ϵ, the empty set by $\emptyset$, the message space (of a signature scheme) by $\mathcal{M}$, an unspecified polynomial by $\mathsf{poly}(\cdot)$, an unspecified negligible function by $\mathsf{negl}(\cdot)$, and the security parameter by λ. PPT stands for probalistic polynomial time. We say that P is a partition of a set $U(\subseteq [n])$ if it satisfies the following conditions: $P = (S_1, \ldots, S_p)$, $p \in [|U|]$, $S_1, \ldots, S_p \in 2^U \setminus \{\emptyset\}$, $\bigcup_{i \in [p]} S_i = U$, and $S_i \cap S_j = \emptyset$ for all $i \neq j$ $(i, j \in [p])$.

2.2 Aggregate Signatures

Here, we recall an ordinary aggregate signature. In this paper, for simplicity, we deal with aggregate signatures which aggregate only one message and signature pair under one verification key.[2]

An aggregate signature scheme consists of the five PPT algorithms (KeyGen, Sign, Verify, Agg, AggVerify).

- $(\mathsf{pk}, \mathsf{sk}) \leftarrow \mathsf{KeyGen}(1^\lambda)$: KeyGen is the key generation algorithm that takes 1^λ as input and outputs a pair $(\mathsf{pk}, \mathsf{sk})$ of public and secret keys.
- $\sigma \leftarrow \mathsf{Sign}(\mathsf{sk}, m)$: Sign is the signing algorithm that takes a secret key sk and a message $m \in \mathcal{M}$ as input, and outputs a signature σ.
- $1 / 0 \leftarrow \mathsf{Verify}(\mathsf{pk}, m, \sigma)$: Verify is the verification algorithm (for an individual signature) that takes a public key pk, a message m, and a signature σ as input, and outputs either 1 (valid) or 0 (invalid).
- $\tau \leftarrow \mathsf{Agg}(\{(\mathsf{pk}_i, m_i, \sigma_i)\}_i)$: Agg is the aggregation algorithm that takes as input tuples of a public key, a message, and a signature, $\{(\mathsf{pk}_i, m_i, \sigma_i)\}_i$. It then outputs an aggregate signature τ.
- $1 / 0 \leftarrow \mathsf{AggVerify}(\{(\mathsf{pk}_i, m_i)\}_i, \tau)$: AggVerify is the verification algorithm (for an aggregate signature) that takes as input pairs of a public key and a message $\{(\mathsf{pk}_i, m_i)\}_i$, and an aggregate signature τ. It then outputs either 1 (valid) or 0 (invalid).

Definition 1 (Correctness). *An aggregate signature scheme* $\Sigma_{\mathrm{AS}} = (\mathsf{KeyGen}, \mathsf{Sign}, \mathsf{Verify}, \mathsf{Agg}, \mathsf{AggVerify})$ *satisfies correctness if for any* $\lambda \in \mathbb{N}$, *any* $n = \mathrm{poly}(\lambda)$, *and any* $m_1, \ldots, m_n \in \mathcal{M}$, *it holds that*

$$\Pr\left[1 \leftarrow \mathsf{AggVerify}(\{(\mathsf{pk}_i, m_i)\}_{i \in [n]}, \tau) \;\middle|\; \begin{array}{l} \forall i \in [n],\ (\mathsf{pk}_i, \mathsf{sk}_i) \leftarrow \mathsf{KeyGen}(1^\lambda), \\ \text{and } \sigma_i \leftarrow \mathsf{Sign}(\mathsf{sk}_i, m_i); \\ \tau \leftarrow \mathsf{Agg}(\{(\mathsf{pk}_i, m_i, \sigma_i)\}_{i \in [n]}) \end{array}\right] = 1.$$

For security, we consider EUF-CMA (Existential UnForgeability against Chosen Message Attacks) security in the model where all key pairs are generated honestly (honest-key model).

Definition 2 (EUF-CMA security). *An aggregate signature scheme* $\Sigma_{\mathrm{AS}} = (\mathsf{KeyGen}, \mathsf{Sign}, \mathsf{Verify}, \mathsf{Agg}, \mathsf{AggVerify})$ *satisfies EUF-CMA security if for any* $\lambda \in \mathbb{N}$, *any* $n = \mathrm{poly}(\lambda)$ *and any PPT adversary* $\mathcal{A}$, *it holds that* $\Pr\left[\mathsf{ExpAS}^{EUF\text{-}CMA}_{\Sigma_{\mathrm{AS}},\mathcal{A}}(\lambda, n) = 1\right] = \mathrm{negl}(\lambda)$ *where* $\mathsf{ExpAS}^{EUF\text{-}CMA}_{\Sigma_{\mathrm{AS}},\mathcal{A}}(\lambda, n)$ *is the following experiment:*

[2] In general, aggregate signatures can aggregate multiple signatures even if they are generated under the same key.

$\mathsf{ExpAS}^{EUF\text{-}CMA}_{\Sigma_{\mathrm{AS}},\mathcal{A}}(\lambda, n)$

$\forall i \in [n], (\mathsf{pk}_i, \mathsf{sk}_i) \leftarrow \Sigma_{\mathrm{AS}}.\mathsf{KeyGen}(1^\lambda);$ $Q := \emptyset;$ $(\{m_i\}_{i \in S}, \tau, S) \leftarrow$ $\mathcal{A}^{\Sigma_{\mathrm{AS}}.\mathsf{Sign}(\mathsf{sk}_1, \cdot)}(\mathsf{pk}_1, \{(\mathsf{pk}_i, \mathsf{sk}_i)\}_{i \in [n] \setminus \{1\}})$	Output 1 if $S \subseteq [n]$, $1 \in S$, $m_1 \notin Q$ and : $\Sigma_{\mathrm{AS}}.\mathsf{AggVerify}(\{(\mathsf{pk}_i, m_i)\}_{i \in S}, \tau) = 1$, else output 0

where when $\mathcal{A}$ makes a query $m \in \mathcal{M}$ to the signing oracle $\Sigma_{\mathrm{AS}}.\mathsf{Sign}(\mathsf{sk}_1, \cdot)$, it computes $\sigma \leftarrow \Sigma_{\mathrm{AS}}.\mathsf{Sign}(\mathsf{sk}_1, m)$, sends σ to $\mathcal{A}$, and sets $Q \leftarrow Q \cup \{m\}$.

Note that in the experiment, the user index 1 is used as a challenge user, whose secret key is unknown to an adversary, and the remaining keys $(\mathsf{pk}_i, \mathsf{sk}_i)_{i \in [n] \setminus \{1\}}$ are directly given to $\mathcal{A}$. Thus, the signing oracle is provided only for the index 1.

2.3 Aggregate Signatures with Interactive Tracing Functionality

Here we recall the formal definitions for an aggregate signature scheme with interactive tracing functionality (ASIT) in the form given in [12].

Definition 3 (ASIT). *An ASIT scheme consists of the PPT algorithms* (KeyGen, Sign, Agg, Verify, PartVerify, Trace) *that work as follows.*

- $(\mathsf{pk}, \mathsf{sk}) \leftarrow \mathsf{KeyGen}(1^\lambda)$: KeyGen *is the key generation algorithm that takes* 1^λ *as input and outputs a pair* $(\mathsf{pk}, \mathsf{sk})$ *of public and secret keys.*
- $\sigma \leftarrow \mathsf{Sign}(\mathsf{sk}, m)$: Sign *is the signing algorithm that takes a secret key* sk *and a message* $m \in \mathcal{M}$ *as input, and outputs a signature* σ.
- $1\ /\ 0 \leftarrow \mathsf{Verify}(\mathsf{pk}, m, \sigma)$: Verify *is the verification algorithm (for an individual signature) that takes a public key* pk, *a message* m, *and a signature* σ *as input, and outputs either* 1 (valid) *or* 0 (invalid).
- $\tau \leftarrow \mathsf{Agg}(f, \{(\mathsf{pk}_i, m_i, \sigma_i)\}_i)$: Agg *is the aggregation algorithm that takes as input a feedback* f *from the verifier and tuples of a public key, a message , and a signature* $\{(\mathsf{pk}_i, m_i, \sigma_i)\}_i$ *as input. It then outputs an aggregate signature* τ.
- $1\ /\ 0 \leftarrow \mathsf{PartVerify}(\beta, \{(\mathsf{pk}_i, m_i)\}_i, \tau, j)$: PartVerify *is the partial verification algorithm that takes as input the verifier's internal state* β, *pairs of a public key and a message* $\{(\mathsf{pk}_i, m_i)\}_i$, *an aggregate signature* τ, *and a target user index* j. *It then outputs either* 1 (valid) *or* 0 (invalid).
- $(\beta', f, V) \leftarrow \mathsf{Trace}(\beta, \{(\mathsf{pk}_i, m_i)\}_i, \tau)$: Trace *is the tracing algorithm that takes as input the verifier's internal state* β, *pairs of a public key and a message* $\{(\mathsf{pk}_i, m_i)\}_i$, *and an aggregate signature* τ. *It then outputs a tuple* (β', f, V) *where* β' *is the verifier's internal state in the next round,* f *is a feedback, and* V *is the traced user set. It is required that the feedback* f *and traced user set* V *can be uniquely retrieved from the internal state* β'.

Trace and PartVerify are required to satisfy the following requirements.

Definition 4 (Correctness of Trace). *Let* Σ_{ASIT} *be an ASIT scheme. The algorithm* $\Sigma_{\mathrm{ASIT}}.\mathsf{Trace}$ *satisfies correctness if for any* $\lambda \in \mathbb{N}$, *any* $n = \mathrm{poly}(\lambda)$, *any* $t \in \mathbb{N}$, *and any* $m_1, \ldots, m_n \in \mathcal{M}$, *it holds that* $\Pr[V_t = \emptyset] = 1$ *where* V_t *is the value in the experiment* $\mathsf{ExpASIT}^{\mathrm{Trace}}_{\Sigma_{\mathrm{ASIT}}}(\lambda, n, \{m_i\}_{i \in [n]})$ *described in Fig. 1 (left).*

$\mathsf{ExpASIT}^{\mathsf{Trace}}_{\Sigma_{\mathrm{ASIT}}}(\lambda, n, \{m_i\}_{i\in[n]})$:

for $i := 1$ **to** n **do**
  $(\mathsf{pk}_i, \mathsf{sk}_i) \leftarrow \Sigma_{\mathrm{ASIT}}.\mathsf{KeyGen}(1^\lambda)$
$t := 1$; $V_1 := \epsilon$; $\beta_1 := \epsilon$; $f_0 := \epsilon$
while true
  for $i := 1$ **to** n **do**
    $\sigma_{i,t} \leftarrow \Sigma_{\mathrm{ASIT}}.\mathsf{Sign}(\mathsf{sk}_i, m_{i,t})$
  $\tau_t \leftarrow$
    $\Sigma_{\mathrm{ASIT}}.\mathsf{Agg}(f_{t-1}, \{(\mathsf{pk}_i, m_{i,t}, \sigma_{i,t})\}_{i\in[n]})$
  $(\beta_{t+1}, f_t, V_t) \leftarrow$
    $\Sigma_{\mathrm{ASIT}}.\mathsf{Trace}(\beta_t, \{(\mathsf{pk}_i, m_{i,t})\}_{i\in[n]}, \tau_t)$
  $t \leftarrow t+1$

$\mathsf{ExpASIT}^{\mathsf{PartVrf}}_{\Sigma_{\mathrm{ASIT}}}(\lambda, n, \beta, j, \{m_i\}_{i\in[n]})$:

for $i := 1$ **to** n **do**
  $(\mathsf{pk}_i, \mathsf{sk}_i) \leftarrow \Sigma_{\mathrm{ASIT}}.\mathsf{KeyGen}(1^\lambda)$
  $\sigma_i \leftarrow \Sigma_{\mathrm{ASIT}}.\mathsf{Sign}(\mathsf{sk}_i, m_i)$
Let f be the feedback and V be the set of traced users that are determined by β
$\tau \leftarrow \Sigma_{\mathrm{ASIT}}.\mathsf{Agg}(f, \{(\mathsf{pk}_i, m_i, \sigma_i)\}_{i\in[n]\setminus V})$
$v \leftarrow$
  $\Sigma_{\mathrm{ASIT}}.\mathsf{PartVerify}(\beta, \{(\mathsf{pk}_i, m_i)\}_{i\in[n]\setminus V}, \tau, j)$
return v

Fig. 1. The experiments used for defining the correctness of an ASIT scheme.

Definition 5 (Correctness of PartVerify). *Let Σ_{ASIT} be an ASIT scheme. The algorithm $\Sigma_{\mathrm{ASIT}}.\mathsf{PartVerify}$ satisfies correctness if for any $\lambda \in \mathbb{N}$, any $n = \mathrm{poly}(\lambda)$, any possible form of internal state β, any $j \in [n]$, and any $m_1, \ldots, m_n \in \mathcal{M}$, it holds that $\Pr[v = 1] = 1$ where v is an output of the experiment $\mathsf{ExpASIT}^{\mathsf{PartVrf}}_{\Sigma_{\mathrm{ASIT}}}(\lambda, n, \beta, j, \{m_i\}_{i\in[n]})$ described in Fig. 1 (right).*

EUF-CMA Security. We recall EUF-CMA security for ASIT. A potential adversary in this security notion includes all signers and an aggregator.

Definition 6. *An ASIT scheme Σ_{ASIT} satisfies EUF-CMA security if for any $\lambda \in \mathbb{N}$, any $n = \mathrm{poly}(\lambda)$, and any PPT adversary $\mathcal{A}$, it holds that $\Pr[\mathsf{ExpASIT}^{\mathrm{EUF\text{-}CMA}}_{\Sigma_{\mathrm{ASIT}},\mathcal{A}}(\lambda, n) = 1] = \mathrm{negl}(\lambda)$ where $\mathsf{ExpASIT}^{\mathrm{EUF\text{-}CMA}}_{\Sigma_{\mathrm{ASIT}},\mathcal{A}}$ is the following experiment.*

$\mathsf{ExpASIT}^{\mathrm{EUF\text{-}CMA}}_{\Sigma_{\mathrm{ASIT}},\mathcal{A}}(\lambda, n)$

$\forall i \in [n], (\mathsf{pk}_i, \mathsf{sk}_i) \leftarrow \Sigma_{\mathrm{ASIT}}.\mathsf{KeyGen}(1^\lambda)$;
$t := 1; Q := \emptyset; W_1 := \emptyset$; : Output 0 when $\mathcal{A}$ halts
run $\mathcal{A}^{O_S(\cdot), O_V(\cdot)}(\mathsf{pk}_1, \{(\mathsf{pk}_i, \mathsf{sk}_i)\}_{i\in[n]\setminus\{1\}})$

where $\mathcal{A}$ can halt at an arbitrary point, $\mathcal{A}$ is allowed to make arbitrarily (polynomially) many queries to the signing oracle O_S and the verification oracle O_V, which work as follows:

O_S: *Given a query $m \in \mathcal{M}$ from $\mathcal{A}$, O_S runs $\sigma \leftarrow \Sigma_{\mathrm{ASIT}}.\mathsf{Sign}(\mathsf{sk}_1, m)$, returns σ to $\mathcal{A}$, and updates $Q \leftarrow Q \cup \{m\}$.*

O_V: *Given pairs of an index and a message $\{(i, m_{i,t})\}_{i\in I_t}$ and an aggregate signature τ from $\mathcal{A}$, O_V outputs 1 (indicating that $\mathcal{A}$ wins) and terminates the experiment if it holds that $\Sigma_{\mathrm{ASIT}}.\mathsf{PartVerify}(\beta_t, \{(\mathsf{pk}_i, m_{i,t})\}_{i\in I_t}, \tau_t, 1) = 1$, $1 \notin W_t$, and $m_{1,t} \notin Q$. Otherwise, O_V executes $(\beta_{t+1}, f_t, V_t) \leftarrow \Sigma_{\mathrm{ASIT}}.\mathsf{Trace}(\beta_t, \{(\mathsf{pk}_i, m_{i,t})\}_{i\in I_t}, \tau_t)$, returns (f_t, V_t) to $\mathcal{A}$, and updates $W_t = W_t \cup V_t$ and $t \leftarrow t+1$.*

Note that the user index 1 is treated as the challenge user, and an adversary is given the secret keys for the remaining users with index 2 to n, and thus the signing oracle is necessary only for the user index 1. Note also that the experiment can output 1 only if $\mathcal{A}$ makes an O_V-query that contains a forged signature with respect to the user index 1 (judged using the algorithm $\mathsf{PartVerify}$).

***R*-Identifiability and Correctness.** We recall R-identifiability and correctness of an ASIT scheme Σ_{ASIT}. R-identifiability guarantees that a verifier can identify all rogue signers within R rounds of executions of the tracing procedure. A potential adversary in these security notions is a set of users $C \subseteq [n]$ that may generate invalid signatures. On the other hand, correctness guarantees that no honest signers will be traced. Note that an aggregator and the verifier behave honestly. These security notions are defined based on the following experiment $\mathsf{ExpASIT}_{\Sigma_{\mathrm{ASIT}},\mathcal{A}}(\lambda, n)$ in which a stateful adversary $\mathcal{A}$ is executed:

$\mathsf{ExpASIT}_{\Sigma_{\mathrm{ASIT}},\mathcal{A}}(\lambda, n)$

$\forall i \in [n], (\mathsf{pk}_i, \mathsf{sk}_i) \leftarrow \Sigma_{\mathrm{ASIT}}.\mathsf{KeyGen}(1^\lambda);$
$C \leftarrow \mathcal{A}(\{(\mathsf{pk}_i, \mathsf{sk}_i)\}_{i\in[n]});$
$t := 1; r := 0; W_1 := \emptyset; \beta_1 := \epsilon;$
$f_0 := \epsilon; I_1 := [n]; J_1 := C;$
run $\mathcal{A}^{O_T(\cdot)}(\{(\mathsf{pk}_i, \mathsf{sk}_i)\}_{i\in[n]})$

: Output $(W := \bigcup_{t'=1}^{t} V_{t'}, C, r)$ when $\mathcal{A}$ halts

where $\mathcal{A}$ can halt at an arbitrary point, and $\mathcal{A}$ is allowed to make arbitrarily (polynomially) many queries to the tracing oracle O_T. Let $W_t := \bigcup_{t'=1}^{t} V_t$, $I_t := [n] \backslash W_t$, and $J_t := C \setminus W_t$. Given a query $(\{m_{i,t}\}_{i\in I_t}, \{(m_{j,t}, \sigma_{j,t})\}_{j\in J_t})$ from $\mathcal{A}$, O_T operates as follows:

1. If there exists $j \in J_t$ such that $\Sigma_{\mathrm{ASIT}}.\mathsf{Verify}(\mathsf{pk}_j, m_{j,t}, \sigma_{j,t}) = 0$, then set $r \leftarrow r + 1$.
2. For every $i \in I_t$, compute $\sigma_{i,t} \leftarrow \Sigma_{\mathrm{ASIT}}.\mathsf{Sign}(\mathsf{sk}_i, m_{i,t})$.
3. Compute $\tau_t \leftarrow \Sigma_{\mathrm{ASIT}}.\mathsf{Agg}(f_{t-1}, \{(\mathsf{pk}_i, m_{i,t}, \sigma_{i,t})\}_{i\in I_t \cup J_t})$.
4. Compute $(\beta_{t+1}, f_t, V_t) \leftarrow \Sigma_{\mathrm{ASIT}}.\mathsf{Trace}(\beta_t, \{(\mathsf{pk}_i, m_{i,t})\}_{i\in I_t \cup J_t}, \tau_t)$.
5. Return (f_t, V_t) to $\mathcal{A}$ and set $t \leftarrow t + 1$.

We define R-identifiability and correctness of ASIT as follows.

Definition 7 (*R*-Identifiability). *An ASIT scheme Σ_{ASIT} satisfies R-identifiability if for any $\lambda \in \mathbb{N}$, any $n = \mathrm{poly}(\lambda)$, and any PPT adversary $\mathcal{A}$, we have*

$$\Pr[(C \not\subseteq W) \mid (W, C, r) \leftarrow \mathsf{ExpASIT}_{\Sigma_{\mathrm{ASIT}},\mathcal{A}}(\lambda, n) \wedge (r \geq R)] = \mathrm{negl}(\lambda).$$

Definition 8 (Correctness). *An ASIT scheme Σ_{ASIT} satisfies correctness if both $\Sigma_{\mathrm{ASIT}}.\mathsf{Trace}$ and $\Sigma_{\mathrm{ASIT}}.\mathsf{PartVerify}$ satisfy correctness, and for any $\lambda \in \mathbb{N}$, any $n = \mathrm{poly}(\lambda)$, and any PPT adversary $\mathcal{A}$, we have*

$$\Pr[([n] \backslash C) \cap W \neq \emptyset \mid (W, C, r) \leftarrow \mathsf{ExpASIT}_{\Sigma_{\mathrm{ASIT}},\mathcal{A}}(\lambda, n)] = \mathrm{negl}(\lambda).$$

3 Feedback Waiting Time in ASIT

This section presents our first contribution. Recall that, as discussed in Sect. 1, the feedback waiting time of a multi-round FT-AS scheme could be critical if we use it in practice. Therefore, here we evaluate the feedback waiting time of a concrete instantiation of ASIT. More specifically, we evaluate the instantiation of ASIT in [12], which is constructed from an ordinary aggregate signature scheme and DTT. In Sect. 3.1, we introduce the concrete instantiation FT-2 of DTT that is proposed by Fiat and Tassa [7], and an ASIT scheme proposed by Ishii et al. [12] that is based on an aggregate signature scheme and FT-2. While Ishii et al. [12] gave the formal definition of DTT, we do not introduce it in this paper due to the space limitation. In Sect. 3.2, we evaluate the feedback waiting time of the ASIT scheme. We remark that this is the first experimental result for ASIT schemes, because [12] only proposed the scheme.

3.1 Existing Instantiations of DTT and ASIT

DTT is a method to trace piracy in a contents distributing service, which works as follows: Users are divided into several subgroups, and for each subgroup a content with a unique watermark is distributed. Once a piracy is found, the contents distributor checks the watermark, divides the corresponding subgroup into smaller subgroup, and repeats this procedure until it traces a piracy. We refer to the interval between the executions of tracing as 1 *round*. Fiat and Tassa [7] proposed two DTTs, and we recall one of them.

Figure 2 illustrates the instantiation of DTT proposed in [7], named FT-2. It is known that FT-2 traces all traitors in $d(\log_2 N + 1)$ rounds, where the number of signers is N and the number of traitors is d.

We recall the ASIT scheme proposed by Ishii et al. [12]. The idea behind the tracing functionality of this scheme is to regard rogue signers as pirates, and run FT-2 to trace them. Figure 3 illustrates the ASIT scheme Σ_{ASIT} based on an aggregate signature scheme Σ_{AS} and the DTT FT-2 Σ_{DTT}. We call this instantiation as AS-FT-2.

3.2 Evaluation of the Feedback Waiting Time of AS-FT-2

We evaluate the feedback waiting time of AS-FT-2, describe the implementation experiment of AS-FT-2 in detail and present the results. We also discuss whether the feedback waiting time meets the requirement for some applications.

Experimental Environment. We use a simulator implemented using Amazon Web Service (AWS) shown in [27]. For the devices, we use four Amazon EC2 instances, which are virtual devices provided by AWS. The EC2 instance types are all t3.micro (virtual CPU: 2, memory: 1 GiB, maximum bandwidth: 5 Gbps). These four instances are managed by Amazon Elastic Kubernetes Service (EKS) as shown in Fig. 4 and each of them has a container image, which consists of alpine Linux 3.15 (latest) as the OS and g++ 10.3.1 as the C++ version.

```
Initialize(1^λ, 1^n) :
  I := {1, ··· , n}; P_1 := I; α_1 := P_1
  return (α_1, P_1)
```

```
Halve(I) :
  A := the first ⌊|I|/2⌋ elements of I; B := I \ A
  return (A, B)
```

```
Trace(α_t, i) :
   P_t := α_t; p := (|P_t| − 1)/2; V := ∅; parse (Q_j)_{j∈[|P|]} ← P_t
   if Q_i = I then
     (L_{p+1}, R_{p+1}) ← Halve(I); I ← ∅; P ← P ∪ ({L_{p+1}} ∪ {R_{p+1}}); p ← p + 1
   else if ∃j ∈ [p] : Q_i = L_j then
 ⋆    I ← I ∪ R_j
 ⋆    if |L_j| = 1 then
 ⋆      V ← L_j; p ← p − 1; P ← P \ ({L_j} ∪ {R_j})
 ⋆    else (L_j, R_j) ← Halve(L_j)
   else if ∃j ∈ [p] : Q_i = R_j then
     Execute the lines marked with ⋆ with switching L_j and R_j.
   P_{t+1} := P; V_t := V; α_{t+1} := P
   return (α_{t+1}, P_{t+1}, V_t)
```

Fig. 2. The descriptions of the DTT (FT-2) by Fiat and Tassa [7] and the subroutine Halve used in the scheme.

Experimental Setting. We capture a situation where some fixed rogue signers among many signers always send invalid signatures. The number N of all signers and the number d of rogue signers are $N = 1000, 3000$, and $d = 5, 10, 40$, respectively. These parameter settings are based on [22], which deals with identifying sensor nodes that have failed to transmit in a sensor network. It assumes a network setting similar to ours where a large number (1000 - 3000) of sensor nodes transmit data and some faulty nodes are included. Each honest/rogue signer sends a pair of a 128-bit message and a signature to the aggregator. For the underlying aggregate signature scheme, we implement the BGLS aggregate signature scheme [3] by using a pairing cryptography library mcl [20] and BN254 [2] for elliptic curves. The experiment is conducted 10 times and the average is taken.

Table 1. Simulation results of AS-FT-2. **Trace** is the average execution time of the tracing algorithm, and **Feedback** is the average time from the time that the aggregator generates an aggregate signature to the time that the verifier returns a feedback.

N	d	**Trace** [ms]	**Feedback** [ms]
1000	5	227.0	233.3
1000	10	171.9	179.6
1000	40	81.2	96.9
3000	5	594.4	605.3
3000	10	452.2	462.1
3000	40	205.2	221.6

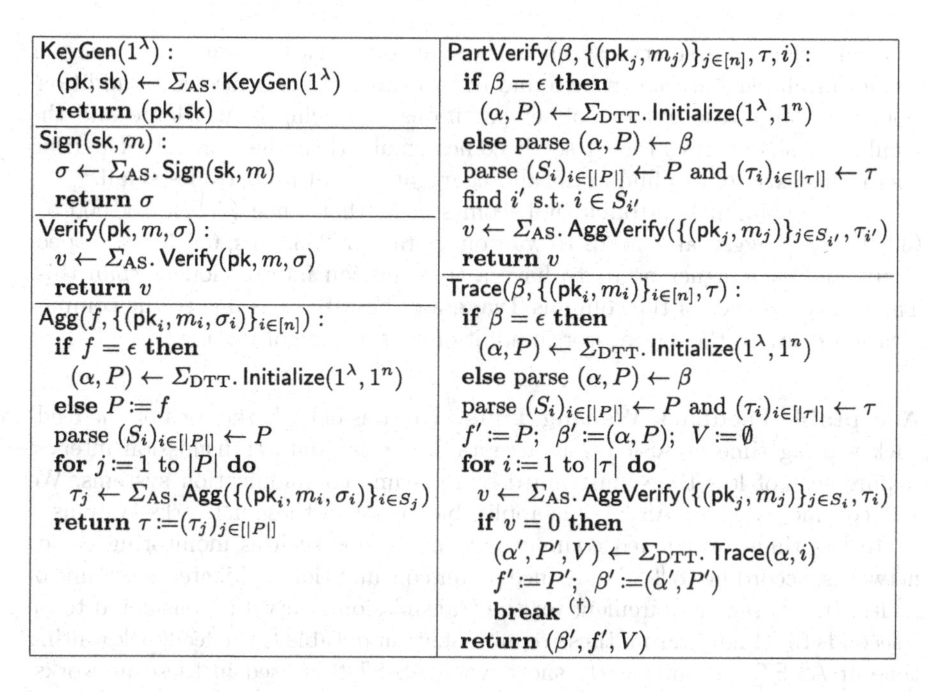

Fig. 3. The instantiation Σ_{ASIT} of an ASIT scheme based on an aggregate signature scheme Σ_{AS} and FT-2 Σ_{DTT} by [12]. (†) This command exits **for** and moves to the next line.

Experimental Results. Table 1 shows the simulation results of AS-FT-2 under the experimental conditions described above. Observe that the execution time of tracing occupies a large ratio of the feedback waiting time, which further increases when N becomes large. This is because the verifier requires $N+1$ pairing operations and a pairing operation takes longer time than other group operations in our instantiation.[3] We observe that the execution time of tracing becomes less as d increases. In this experiment, each rogue signer always sends

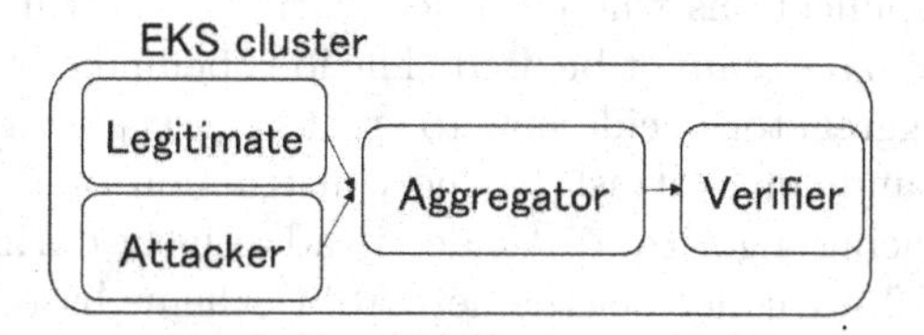

Fig. 4. Device configuration in a cluster. The legitimate node sends $N-d$ valid signatures and the rogue signer node sends d invalid signatures in each round, where N is the number of all signers and d is the number of rogue signers.

[3] It takes 0.679 ms per group operation and 286 ms per pairing operation in our setting.

an invalid signature, so at least one of the splits out of two includes a rogue signer and an invalid signature can be found in two times of verifications. In addition, when d is large, the number of set partitions in tracing is also large and the number of signatures in each set becomes small, then the number of pairing operations executed to find an invalid aggregate signature becomes small.[4]

For the column **Feedback**, it becomes clear that when $(N, d) = (1000, 5)$, $(3000, 5)$, the aggregator needs to wait on average of 233.3 ms, 605.3 ms, respectively, and each signer needs to have a transmission interval longer than this. The time difference of the columns **Trace** and **Feedback** is due to the communication delay to the aggregator in addition to the time of tracing.

Acceptable Feedback Waiting Time. We consider whether or not the feedback waiting time of AS-FT-2 is acceptable for the data transmission interval requirement of less time-constrained or real-time communication systems. We also consider whether AS-FT-2 is applicable or not in these networks systems.

In less time-constrained communication systems such as monitoring sensor networks, according to [28], to reduce power consumption and increase lifetime of nodes, the average requirement for the transmission interval is considered to be 5 seconds to an hour. From these requirements and Table 1, the feedback waiting time of AS-FT-2 is sufficiently short when AS-FT-2 is used in these networks. Therefore, AS-FT-2 is applicable in monitoring sensor networks and industrial networks.

On the other hand, in some real-time communication systems, according to [29], to guarantee traffic safety under bandwidth constraints, the data transmission interval for a cooperative intelligent transportation systems is considered to be 200 ms. From these requirements and Table 1, the feedback waiting time of AS-FT-2 is too long. Therefore, AS-FT-2 is *not* applicable for such real-time communication systems.

4 An ASIT Scheme Without Feedback

In this section, we propose a new ASIT scheme that does not require a feedback, in anticipation of applications where the feedback waiting time is short so that the ASIT scheme AS-AT-2 cannot be used. The idea behind our proposed scheme is that we let the aggregator decide how to create aggregate signatures (i.e., the partitions) beforehand, in contrast to the construction of AS-FT-2, in which the verifier (the tracing algorithm) decides it adaptively during the execution. Similarly to AS-FT-2, we construct the new ASIT scheme based on an aggregate signature scheme and a piracy tracing algorithm, but in our scheme we use a Sequential Traitor Tracing (STT) [23]. STT predetermines the assignment of watermarks beforehand, and this feature allows us to remove a feedback from an ASIT scheme.

[4] In the tracing algorithm of AS-FT-2, when the verifier finds an invalid aggregate signature, it outputs a feedback and the rest of signatures are not verified.

We first review the concept of STT, and then the concrete instantiation of STT, named c-SW-1 [23]. Then we provide our new construction of ASIT named AS-SW-1 based on an aggregate signature scheme and c-SW-1. As explained earlier, [12] proposed a generic construction of ASIT based on an aggregate signature scheme and a DTT. Our ASIT scheme AS-SW-1 is constructed just by replacing FT-2 used in AS-FT-2 with c-SW-1. Therefore, to prove AS-SW-1 satisfies the requirements of an ASIT scheme, it is sufficient to prove that c-SW-1 is a DTT. Since this is rather straightforward to see, due to the space limitation, we omit the formal proof for this fact.

4.1 Sequential Traitor Tracing

As already mentioned, STT is a variant of a piracy tracing algorithm that predetermines the watermark assignment. The assignment is actually determined by an allocation matrix M, whose entry is decided by a function family Φ. The column of M corresponds to the allocation of watermarks for each distribution. The time interval between each distribution is called *segment.* In this subsection, we introduce the allocation matrix and the function family along with several lemmas and an implementation of the function family. Then, we provide an implementation of the STT.

Let $W := [q]$ be the set of watermarks, and b $(\leq q)$ and l be integers. We let $\Phi = \{\phi_{i,j} : 1 \leq i \leq b, 1 \leq i \leq l\}$, where $\phi_{i,j} : W \to W$ is a function family that satisfies the following conditions:

Condition 1. *For all $x \in W$, any fixed j and a pair of the first indexes (i_1, i_2) where $i_1 \neq i_2$, it holds that $\phi_{i_1 j}(x) \neq \phi_{i_2 j}(x)$.*

Condition 2. *For any (i_1, i_2) and (j_1, j_2) where $j_1 \neq j_2$, and any $x_1, x_2 \in W$ such that $x_1 \neq x_2$, if $\phi_{i_1,j_1}(x_1) = \phi_{i_2,j_1}(x_2)$ then $\phi_{i_1,j_2}(x_1) \neq \phi_{i_2,j_2}(x_2)$.*

Now, we describe an allocation matrix M. Let Φ be a function family that satisfies the above conditions. Let M_0 and $\phi_{i,j}(M_0)$ $(i \in [b],\ j \in [l])$ be the following $q \times 1$ matrices:

$$M_0 = \begin{pmatrix} 1 \\ 2 \\ \vdots \\ q \end{pmatrix}, \phi_{i,j}(M_0) = \begin{pmatrix} \phi_{i,j}(1) \\ \phi_{i,j}(2) \\ \vdots \\ \phi_{i,j}(q) \end{pmatrix}.$$

Then, the allocation matrix M is described as follows:

$$M = \begin{pmatrix} M_0 \ \phi_{1,1}(M_0) \ \phi_{1,2}(M_0) \cdots \phi_{1,l}(M_0) \\ M_0 \ \phi_{2,1}(M_0) \ \phi_{2,2}(M_0) \cdots \phi_{2,l}(M_0) \\ \vdots \quad \vdots \quad \vdots \quad \vdots \\ M_0 \ \phi_{b,1}(M_0) \ \phi_{b,2}(M_0) \cdots \phi_{b,l}(M_0) \end{pmatrix}.$$

Note that M has b block rows and each block consists of q rows. The i-th row of M represents the watermark assigned to user i in order (thus, it is implicitly assumed that the number of users is bq), and j-th column represents the assignment of watermarks for each user at segment j. Let (r,k) denote the k-th row of the r-th block.

Here we explain how M is used to detect piracy in a content distribution service. The content distributor distributes a watermark of a content $M(i,j)$ to the user i at the segment j, with monitoring piracy. For ease of discussion, we assume that the distributor detects exactly one piracy in every segment. Let $F_j = (f_1, \dots, f_j)$ be the index sequence of the watermarks on the content detected by the distributor from segment 1 to segment j. Furthermore, let

$$\rho(F_j, i, s) = \begin{cases} 1 \text{ (if } f_s = M(i,j)) \\ 0 \text{ (otherwise)} \end{cases}, \rho(F_j, i) = \sum_{s=1}^{j} \rho(F_j, i, s).$$

If there exists a user i such that $\rho(F_j, i) \geq t$ for a threshold t, then the distributor regards i as a pirate and eliminates it. Regarding the threshold, Safavi-Naini and Wang [23] showed the following lemmas.

Lemma 1 ([23]). *Let C be the set of pirates where $|C| = c$, and $F_j = (f_1, \dots, f_j)$ be the index sequence of the watermarks on the content detected by the distributor from segment 1 to segment j. If there exists a user i such that $\rho(F_j, i) \geq c+1$, then $i \in C$.*

Regarding how many segments are necessary to trace all pirates, they also showed the following lemma.

Lemma 2 ([23]). *Let C be the set of pirates where $|C| = c$. All pirates can be traced at most $c^2 + c$ segments.*

Lemma 2 tells the relation between the number of segments and the number of pirates.

Corollary 1 ([23]). *Let M be an $n \times (l+1)$ allocation matrix. The STT that employs this matrix can trace at most*

$$c = \left\lfloor \frac{-1 + \sqrt{5 + 4l}}{2} \right\rfloor \tag{1}$$

pirates.

Implementation of Φ. Safavi-Naini and Wang [23] proposed the following implementation Φ of the function family.

Theorem 1 ([23]). *Let p be a prime number. Let $\Phi = \{\phi_{i,j} : i, j \in [(p-1)/2]\}$ $(\phi_{i,j} : Z_p^* \to Z_p^*)$ be the function family defined by $\phi_{i,j}(x) = (i+j)x \mod p$. Then, Φ satisfies Condition 1 and Condition 2.*

```
GenMatrix(n, c):
  select a minimum prime p s.t.
   p ≥ max(1 + √(2n), 2c^2 + 2c + 1);
  l:=(p − 1)/2; q = p − 1; b:=⌈n/q⌉
  for j = 1 to l do
    for i = 1 to b do
      for x = 1 to q do
        M[b(i − 1) + x, j] = φ_{i,j}(x)
  return M
```

```
Initialize(1^λ, 1^n):
  j:=1; F:=();
  M ← GenMatrix(n, c) based on Φ
  for x:=1 to q do
    Q_{1,x} :={i ∈ [n] : M[i, 1] = x}
  P_1 :=(Q_{1,1}, ..., Q_{1,q}); α_1 :=(j, F, P_1, M)
  return (α_1, P_1)
```

```
Trace(α, i):
  parse (j, F, P, M) ← α; V:=∅;
  append i to F
  for u = 1 to n do
    if ρ(F, u) = c + 1 then
      V ← u
  for x:=1 to q do
    Q_{j+1,x} :={i ∈ [n] : M[i, j + 1] = x}
  P_{j+1} :=(Q_{j+1,1}, ..., Q_{j+1,q}); j ← j + 1
  α_j :=(j, F, P_j, M)
  return (α_j, P_j, V)
```

Fig. 5. The construction c-SW-1 of STT. GenMatrix is a subroutine that is used to generate the allocation matrix M based on the function family Φ in Theorem 1.

Observation. When we use Φ in Theorem 1 in a STT, the allocation matrix should be the $n \times l$ matrix where $n = (p-1)^2/2$ and $l = (p-1)/2$. Also, the maximum number of pirates that can be traced is $c = \lfloor \frac{-1+\sqrt{2p+3}}{2} \rfloor$. Given n and c, the condition for p is expressed as $p \geq \max(1+\sqrt{2n}, 2c^2+2c-1)$ from $(p-1)^2/2 \geq n$ and $c^2+c \leq (p-1)/2+1$ (note that the second inequality is due to Lemma 2).

An Instantiation of STT. Here we provide an instantiation of STT based on the function family in Theorem 1. Let $n := bq$ be the number of signers, $C \subseteq [n]$ be a set of pirates where $|C| = c \leq n$, and $W := [q]$ be a set of watermarks that are assumed to be given to the distributor in advance. Let $Q_{j,x}$ be the set of users to which the watermark $x \in W$ is assigned at segment j, i.e., $Q_{j,x} = \{i \in [n] : M[i,j] = x\}$.

Figure 5 illustrates the construction c-SW-1 of the STT that uses the function family Φ in Theorem 1. Note that in the subroutine GenMatrix, the prime p is chosen based on the above observation. We can show the following lemma (for which we omit the proof due to the space limitation).

Lemma 3. *The algorithm c-SW-1 is a DTT.*

4.2 The Construction of an ASIT Scheme Without a Feedback

We demonstrate an ASIT scheme that does not require a feedback based on an aggregate signature and the STT c-SW-1 (hereinafter referred to as AS-SW-1).

$\mathsf{KeyGen}(1^\lambda)$:
  $(\mathsf{pk}, \mathsf{sk}) \leftarrow \Sigma_{\mathsf{AS}}.\mathsf{KeyGen}(1^\lambda)$
  return $(\mathsf{pk}, \mathsf{sk})$

$\mathsf{Sign}(\mathsf{sk}, m)$:
  $\sigma \leftarrow \Sigma_{\mathsf{AS}}.\mathsf{Sign}(\mathsf{sk}, m)$
  return σ

$\mathsf{Verify}(\mathsf{pk}, m, \sigma)$:
  $v \leftarrow \Sigma_{\mathsf{AS}}.\mathsf{Verify}(\mathsf{pk}, m, \sigma)$
  return v

$\mathsf{Agg}(\alpha, \{(\mathsf{pk}_i, m_i, \sigma_i)\}_{i \in [n]})$:
  if $\alpha = \epsilon$ **then**
    $j := 1$; $V := \emptyset$
    $(\alpha, P) \leftarrow c\text{-}\mathsf{SW}\text{-}1.\mathsf{Initialize}(1^\lambda, 1^n)$
  parse $(S_i)_{i \in [|P|]} \leftarrow P$
  parse $(j, F, P, M) \leftarrow \alpha$
  for $p := 1$ to $|P|$ **do**
    $\tau_p \leftarrow \Sigma_{\mathsf{AS}}.\mathsf{Agg}(\{(\mathsf{pk}_i, m_i, \sigma_i)\}_{i \in S_p \setminus V})$
  for $x := 1$ to q **do**
    $Q_{j+1,x} := \{i \in [n] \ : \ M[i, j] = x\}$
  $P \leftarrow (Q_{j+1,x})_{x \in [q]}$
  $j \leftarrow j + 1$; $\alpha \leftarrow (j, F, P, M)$
  return $\tau := (\tau_p)_{p \in [|P|]}$

$\mathsf{PartVerify}(\beta, \{(\mathsf{pk}_j, m_j)\}_{j \in [n]}, \tau, i)$:
  if $\beta = \epsilon$ **then**
    $j := 1$; $V := \emptyset$; $(\alpha, P) \leftarrow c\text{-}\mathsf{SW}\text{-}1.\mathsf{Initialize}(1^\lambda, 1^n)$
  else parse $(\alpha, P) \leftarrow \beta$
  parse $(S_i)_{i \in [|P|]} \leftarrow P$ and $(\tau_i)_{i \in [|\tau|]} \leftarrow \tau$
  find i' s.t. $i \in S_{i'}$
  $v \leftarrow \Sigma_{\mathsf{AS}}.\mathsf{AggVerify}(\{(\mathsf{pk}_j, m_j)\}_{j \in S_{i'} \setminus V}, \tau_{i'})$
  return v

$\mathsf{Trace}(\beta, \{(\mathsf{pk}_i, m_i)\}_{i \in [n]}, \tau)$:
  if $\beta = \epsilon$ **then**
    $j := 1$; $V := \emptyset$
    $(\alpha, P) \leftarrow c\text{-}\mathsf{SW}\text{-}1.\mathsf{Initialize}(1^\lambda, 1^n)$
  else parse $(\alpha, P) \leftarrow \beta$
  parse $(S_i)_{i \in [|P|]} \leftarrow P$ and $(\tau_i)_{i \in [|\tau|]} \leftarrow \tau$
  $f := \epsilon$; $\beta' := (\alpha, P)$
  for $i := 1$ to $|\tau|$ **do**
    $v \leftarrow \Sigma_{\mathsf{AS}}.\mathsf{AggVerify}(\{(\mathsf{pk}_j, m_j)\}_{j \in S_i \setminus V}, \tau_i)$
    if $v = 0$ **then**
      $(\alpha', P', V) \leftarrow c\text{-}\mathsf{SW}\text{-}1.\mathsf{Trace}(\alpha, i)$
      $\beta' := (\alpha', P')$
      break (†)
  return (β', f, V)

Fig. 6. AS-SW-1. (†) This command exits **for** and moves to the next line.

Figure 6 illustrates AS-SW-1, where c satisfies the Eq. (1). We assume without loss of generality that the number of users n and the number of rogue signers c are given to the algorithms. Observe that both Agg and Trace run c-SW-1.Initialize when they are initiated. Because c-SW-1, including the subroutine GenMatrix, is deterministic, these algorithms share the same partition P (or the same allocation matrix). Note that the output f by Trace is an empty string. (This feedback is put just to follow the syntax of ASIT, and thus Agg does not receive this f in an actual execution.) Also, when the verifier identifies rogue signers, it sends the traced attacker set V to the aggregator.

Security. Observe that AS-SW-1 is constructed just replacing the DTT FT-2 used in AS-FT-2 with c-SW-1. Furthermore, as shown in Lemma 3, c-SW-1 is a DTT. Therefore, AS-SW-1 follows the generic construction of ASIT proposed by Ishii et al. [12], which indicates the following lemma.

Lemma 4. *Assume the underlying aggregate signature scheme Σ_{AS} is EUF-CMA secure, and the underlying DTT is c-SW-1. Then, AS-SW-1 is an ASIT scheme satisfying EUF-CMA security, $(c^2 + c)$-identifiabilty, and correctness.*

Table 2. Theoretical evaluations. N and d are the numbers of all signers and rogue signers, respectively, and p is a prime number that satisfies $p > \max(2d^2 + 2d, \lceil \frac{\sqrt{2N}}{2} \rceil)$. $R_{\max}$ is the maximum number of rounds required to trace all the rogue signers, and $\mathrm{Sig}_{\max}$ is the maximum number of signatures sent by the aggregator per round.

	$R_{\max}$	$\mathrm{Sig}_{\max}$
AS-FT-2	$d \log_2 N + d$	$2d + 1$
AS-SW-1	$d^2 + d$	$p - 1$

5 Comparison of AS-SW-1 with AS-FT-2

In this section, we theoretically and experimentally evaluate the efficiency of AS-SW-1 compared to that of AS-FT-2. Then, we discuss which of the two schemes is more suitable in some situations.

5.1 Theoretical Evaluations

Given the number N of all signers and the number d of rogue signers, we compare the efficiency of the tracing functionality and the communication bandwidth of AS-FT-2 and AS-SW-1. To compare these metrics, we evaluate the maximum number of rounds required to trace all the rogue signers and the maximum number of signatures sent by the aggregator per round.

From Table 2, for the number of rounds $R_{\max}$, when $d < \log_2 N$, AS-SW-1 requires fewer rounds than AS-FT-2, otherwise AS-FT-2 requires fewer rounds than AS-SW-1. For $\mathrm{Sig}_{\max}$, when $2(d^2 + d) < \frac{\sqrt{2N}}{2}$ and $\frac{\sqrt{2N}}{2} - 1 < 2d + 1$, i.e., $18 \leq N \leq 31$ and $d = 1$, the maximum number of signatures sent by the aggregator per round of AS-SW-1 is less than that of AS-FT-2, otherwise AS-FT-2 sends less signatures than AS-SW-1. Therefore, AS-SW-1 traces more efficiently when the number of rogue signers is quite much smaller compared to the number of all signers but requires more bandwidth than AS-FT-2.

5.2 Implementation Evaluations

Based on the theoretical evaluations, we concretely set the number of all signers and that of the rogue signers, and implement our scheme, and evaluate which of AS-FT-2 or AS-SW-1 is more effective based on experiments. The details of the experimental environment are the same as in Sect. 3.

Simulation Settings. In the experiment setting, we capture a situation where a small number of fixed attackers always send invalid signatures to show the tracing efficiency and bandwidth consumption of AS-SW-1 by comparing with AS-FT-2. As in Sect. 3, the number N of all signers and the number d of rogue signers are $N = 1000$, 3000, and $d = 5, 10, 40$, respectively. The experiment is performed 10 times and the average is taken.

Table 3. Simulation results for AS-FT-2 and AS-SW-1. **non-FB** stands for no feedback ("✓" means it does not require a feedback, and "–" means it does), **Time** is the total time from the start of the first round to the completion of tracing, and **Sig** is the average bytes of transmitted signatures per round.

	N	d	p	**non-FB**	**Time** [s]	**Sig** [kB]
AS-FT-2	1000	5	–	–	87.3	0.42
	1000	10	–	–	150.5	0.61
	1000	40	–	–	490.3	1.54
	3000	5	–	–	270.9	0.42
	3000	10	–	–	498.4	0.59
	3000	40	–	–	1676.7	1.62
AS-SW-1	1000	5	61	✓	58.4	4.70
	1000	10	223	✓	230.4	17.41
	1000	40	3299	✓	3332.3	78.03
	3000	5	79	✓	138.7	6.11
	3000	10	223	✓	560.8	17.39
	3000	40	3299	✓	6912.5	234.72

Simulation Results. Table 3 shows our result. In this table, the prime p is the smallest one satisfying Table 2 for AS-SW-1. The most important part is the column **non-FB**, which indicates the necessity of a feedback. We confirm that AS-SW-1, which does not require a feedback, indeed works well. For tracing efficiency, the total time of tracing is almost proportional to d^2 and N/p for AS-SW-1 when $p < N$, while AS-FT-2 is almost proportional to $d \log N$. This result is consistent with the theoretical evaluation shown in Table 2. From Table 3, the total time of tracing of AS-SW-1 is less than that of AS-FT-2 when $200d < N$, otherwise that of AS-FT-2 is less. On the other hand, for bandwidth consumption, from Table 2, the column **Sig** is almost proportional to d^2 for AS-SW-1 when $p < N$ and otherwise it is the same as individual signature transmission, while AS-FT-2 is almost proportional to d. Specifically, AS-SW-1 requires 144.9 times more signatures to trace rogue signers than AS-FT-2 when $(N, d) = (3000, 40)$. Observe that this result indicates that AS-SW-1 requires more bandwidth consumption than AS-FT-2.

Suitable Schemes for Applications. We consider which of AS-FT-2 and AS-SW-1 is more suitable for some applications, which are less time-constrained or real-time communication systems. In less time-constrained communication systems, e.g., in monitoring sensor networks [28], a system has low bandwidth and could have many rogue devices, say, from 40 to 300 rogue devices out of 1000 devices [22]. From the above results, AS-FT-2 is more efficient than AS-SW-1 in terms of the bandwidth and the time to trace all rogue signers when d is large. For bandwidth, LoRa, a communication protocol used in IoT networks, has a speed of 37.5 kbps [16], which is sufficiently fast for AS-FT-2. Although the aggregator of AS-FT-2 needs to wait for a feedback, as mentioned in Sect. 3, the waiting time is not too long for these applications. Therefore, AS-FT-2 is more suitable for less time-constrained communication systems to take advantage of its bandwidth efficiency and short tracing time when d is large. In some real-time communication systems, e.g., cooperative intelligent transportation systems [29], many devices send data in short interval, which is about 200 ms, so AS-FT-2 cannot be used because of the feedback waiting time. On the other hand, the above results show that AS-SW-1 is more efficient in terms of requiring no feedback and the time to trace all rogue signers when d is small ($d \leq 10$). However, when d is large, AS-SW-1 needs more tracing time and large bandwidth. Therefore, AS-SW-1 is more suitable for real-time communication systems to take advantage of its no feedback feature and short tracing time when d is small.

6 Conclusion

In this paper, to examine whether an existing FT-AS scheme is capable of transmitting a feedback sufficiently fast on a real system, we evaluated the feedback waiting time of the implementation AS-FT-2 of an ASIT scheme proposed in [12]. The results of the implementation experiment indicates that if a system whose acceptable feedback time is significantly larger than a few hundred ms, e.g., industrial sensor systems [14], AS-FT-2 can be used without problems. On the other hand, there may be some applications where such sufficiently large feedback time is not acceptable, e.g., cooperative intelligent transportation systems [29]. In anticipation of such applications, we also proposed an ASIT scheme AS-SW-1 that does not require a feedback. The proposed scheme eliminates a feedback by using the STT proposed by Safavi-Naini et al. [23] in Ishii et al.'s generic construction of ASIT [12]. From our implementation results, we found that although a feedback is completely eliminated in our scheme, its communication cost is 144.9 times larger when $(N, d) = (3000, 40)$. Therefore, AS-SW-1 is more suitable for applications in which it is highly time-constrained and high bandwidth is available, while for applications where bandwidth is constrained but time is not, AS-FT-2 is more suitable.

We leave it as a future work to construct a concrete system that uses ASIT schemes. Towards this goal, we will first need to consider an execution environment such as the network topology and the message format for ASIT schemes.

Acknowledgement. This work was partially supported by the Cabinet Office (CAO), Cross-ministerial Strategic Innovation Promotion Program (SIP), "Cyber Physical Security for IoT Society", JPNP18015 (funding agency: NEDO).

References

1. Ahn, J.H., Green, M., Hohenberger, S.: Synchronized aggregate signatures: new definitions, constructions and applications. In: CCS 2010, pp. 473–484. ACM (2010)
2. Barreto, P.S.L.M., Naehrig, M.: Pairing-friendly elliptic curves of prime order. In: Preneel, B., Tavares, S. (eds.) SAC 2005. LNCS, vol. 3897, pp. 319–331. Springer, Heidelberg (2006). https://doi.org/10.1007/11693383_22
3. Boneh, D., Gentry, C., Lynn, B., Shacham, H.: Aggregate and verifiably encrypted signatures from bilinear maps. In: Biham, E. (ed.) EUROCRYPT 2003. LNCS, vol. 2656, pp. 416–432. Springer, Heidelberg (2003). https://doi.org/10.1007/3-540-39200-9_26
4. Boneh, D., Lynn, B., Shacham, H.: Short signatures from the Weil pairing. In: ASIACRYPT 2001 (2001)
5. Du, D., Hwang, F.K., Hwang, F.: Combinatorial Group Testing and its Applications, vol. 12. World Scientific (2000)
6. Eppstein, D., Goodrich, M.T., Hirschberg, D.S.: Improved combinatorial group testing algorithms for real-world problem sizes. SIAM J. Comput. **36**(5), 1360–1375 (2007)
7. Fiat, A., Tassa, T.: Dynamic traitor tracing. In: Wiener, M. (ed.) CRYPTO 1999. LNCS, vol. 1666, pp. 354–371. Springer, Heidelberg (1999). https://doi.org/10.1007/3-540-48405-1_23
8. Gentry, C., Ramzan, Z.: Identity-based aggregate signatures. In: Yung, M., Dodis, Y., Kiayias, A., Malkin, T. (eds.) PKC 2006. LNCS, vol. 3958, pp. 257–273. Springer, Heidelberg (2006). https://doi.org/10.1007/11745853_17
9. Gerbush, M., Lewko, A., O'Neill, A., Waters, B.: Dual form signatures: an approach for proving security from static assumptions. In: Wang, X., Sako, K. (eds.) ASIACRYPT 2012. LNCS, vol. 7658, pp. 25–42. Springer, Heidelberg (2012). https://doi.org/10.1007/978-3-642-34961-4_4
10. Hartung, G., Kaidel, B., Koch, A., Koch, J., Rupp, A.: Fault-tolerant aggregate signatures. In: Cheng, C.-M., Chung, K.-M., Persiano, G., Yang, B.-Y. (eds.) PKC 2016. LNCS, vol. 9614, pp. 331–356. Springer, Heidelberg (2016). https://doi.org/10.1007/978-3-662-49384-7_13
11. Hohenberger, S., Sahai, A., Waters, B.: Full domain hash from (Leveled) multilinear maps and identity-based aggregate signatures. In: Canetti, R., Garay, J.A. (eds.) CRYPTO 2013. LNCS, vol. 8042, pp. 494–512. Springer, Heidelberg (2013). https://doi.org/10.1007/978-3-642-40041-4_27
12. Ishii, R., et al.: Aggregate signature with traceability of devices dynamically generating invalid signatures. In: Zhou, J., et al. (eds.) ACNS 2021. LNCS, vol. 12809, pp. 378–396. Springer, Cham (2021). https://doi.org/10.1007/978-3-030-81645-2_22
13. Kautz, W.H., Singleton, R.C.: Nonrandom binary superimposed codes. IEEE Trans. Inf. Theory **10**(4), 363–377 (1964)
14. Kiran, M.P.R.S., Rajalakshmi, P.: Performance analysis of CSMA/CA and PCA for time critical industrial IoT applications. IEEE Trans. Ind. Inform. **14**(5), 2281–2293 (2018)
15. Kumar, R., Rajagopalan, S., Sahai, A.: Coding constructions for blacklisting problems without computational assumptions. In: Wiener, M. (ed.) CRYPTO 1999. LNCS, vol. 1666, pp. 609–623. Springer, Heidelberg (1999). https://doi.org/10.1007/3-540-48405-1_38
16. Lavric, A., Popa, V.: Performance evaluation of Lora wan communication scalability in large-scale wireless sensor networks. In: Wireless Communications and Mobile Computing 2018 (2018)

17. Lee, K., Lee, D.H., Yung, M.: Sequential aggregate signatures with short public keys without random oracles. Theor. Comput. Sci. **579**, 100–125 (2015)
18. Lu, S., Ostrovsky, R., Sahai, A., Shacham, H., Waters, B.: Sequential aggregate signatures and multisignatures without random oracles. In: Vaudenay, S. (ed.) EUROCRYPT 2006. LNCS, vol. 4004, pp. 465–485. Springer, Heidelberg (2006). https://doi.org/10.1007/11761679_28
19. Lysyanskaya, A., Micali, S., Reyzin, L., Shacham, H.: Sequential aggregate signatures from trapdoor permutations. In: Cachin, C., Camenisch, J.L. (eds.) EUROCRYPT 2004. LNCS, vol. 3027, pp. 74–90. Springer, Heidelberg (2004). https://doi.org/10.1007/978-3-540-24676-3_5
20. Mitsunari, S.: mcl - a portable and fast pairing-based cryptography library (2016). https://github.com/herumi/mcl
21. Neven, G.: Efficient sequential aggregate signed data. In: Smart, N. (ed.) EUROCRYPT 2008. LNCS, vol. 4965, pp. 52–69. Springer, Heidelberg (2008). https://doi.org/10.1007/978-3-540-78967-3_4
22. Pandey, M., Dhanoriya, S., Bhagat, A.: Fast and efficient data acquisition in radiation affected large WSN by predicting transfaulty nodes. In: Bhattacharyya, P., Sastry, H., Marriboyina, V., Sharma, R. (eds.) Smart and Innovative Trends in Next Generation Computing Technologies. NGCT 2017. Communications in Computer and Information Science, vol. 828, pp. 246–262. Springer, Singapore (2017). https://doi.org/10.1007/978-981-10-8660-1_19
23. Safavi-Naini, R., Wang, Y.: Sequential traitor tracing. IEEE Trans. Inf. Theory **49**(5), 1319–1326 (2003)
24. Sato, S., Shikata, J.: Interactive aggregate message authentication equipped with detecting functionality from adaptive group testing. IACR Cryptol. ePrint Arch. **2020**, 1218 (2020)
25. Sato, S., Shikata, J., Matsumoto, T.: Aggregate signature with detecting functionality from group testing. IACR Cryptol. ePrint Arch. **2020**, 1219 (2020)
26. Shikata, J., Matsumoto, T.: ECSEC: Digital signature system and digital signature method (2021). JP 2021-077961, A, 2021-5-20. (in Japanese)
27. Song, Z., Anzai, R., Sakamoto, J., Yoshida, N., Matsumoto, T.: Proposal and prototype implementation of a cloud-based simulator for traceable aggregate signature protocol. In: SCIS 2022 (2022). (in Japanese)
28. Suryavansh, S., Benna, A., Guest, C., Chaterji, S.: A data-driven approach to increasing the lifetime of IoT sensor nodes. Sci. Rep. **11**(1), 1–12 (2021)
29. Tak, S., Choi, S.: Safety monitoring system of CAVs considering the trade-off between sampling interval and data reliability. Sensors **22**(10), 3611 (2022)
30. Zaverucha, G.M., Stinson, D.R.: Group testing and batch verification. In: Kurosawa, K. (ed.) ICITS 2009. LNCS, vol. 5973, pp. 140–157. Springer, Heidelberg (2010). https://doi.org/10.1007/978-3-642-14496-7_12

Post-quantum Secure Communication with IoT Devices Using Kyber and SRAM Behavioral and Physical Unclonable Functions (Extended Abstract)

Roberto Román(✉), Rosario Arjona, and Iluminada Baturone

Instituto de Microelectrónica de Sevilla, Universidad de Sevilla, CSIC, Seville, Spain
{roman,arjona,lumi}@imse-cnm.csic.es

Abstract. For a secure Internet-of-Things (IoT) ecosystem, not only the establishment of secure communication channels but also the authentication of devices is crucial. Authenticated key exchange protocols establish shared cryptographic keys between the parties and, in addition, authenticate their identities. Usually, the identities are based on a pair of private and public keys. Physical Unclonable Functions (PUFs) are widely used recently to bind physically the private key to a device. However, since PUFs are vulnerable to attacks, even non-invasive attacks without accessing the device, this paper proposes the use of Behavioral and Physical Unclonable Functions (BPUFs), which allow multimodal authentication and are more difficult to be virtually or physically cloned. In order to resist attacks from classic and quantum computers, this paper considers a Kyber key exchange protocol. Recently, Kyber has been selected by the Post-Quantum Cryptography standardization process of the National Institute of Standards and Technology (NIST) for key establishment protocols. In this work, we propose to strengthen a Kyber key exchange protocol with BPUFs extracted from SRAMs included in IoT devices. Experimental results prove the feasibility of the proposal in WiPy boards.

Keywords: Post-quantum authenticated key exchange protocols · Behavioral and Physical Unclonable Functions (BPUFs) · Internet-of-Things (IoT) devices

1 Introduction

The establishment of secure communication channels between authenticated IoT devices is essential in an IoT application. A communication channel between two devices is secure if both devices authenticate each other and share a random session cryptographic key. Typically, Elliptic-Curve Cryptography (ECC) and RSA based asymmetric key algorithms are used for authentication and key exchange protocols, the first ones mostly in the case of devices with constrained resources [1].

To ensure the authenticity of a device, its unique identity is represented by a unique private key. The private key is obtained from a root key stored securely in the device. In the

© The Author(s), under exclusive license to Springer Nature Switzerland AG 2022
W. Li et al. (Eds.): ADIoT 2022, LNCS 13745, pp. 72–83, 2022.
https://doi.org/10.1007/978-3-031-21311-3_9

case of low-cost devices, a good solution is not to store the root key but reconstruct it with the use of a Physical Unclonable Function (PUF), which exploits the unique properties of the device hardware due to the manufacturing process variations. Particularly, SRAM PUFs are widely employed because SRAMs are already included in devices and no additional hardware is required. In the other side, each unique private key is associated with a public key that is the public identity of the device. A Public Key Infrastructure (PKI) like X.509 certificates should be employed to ensure the authenticity of the public keys.

In this paper, we address two threats against the above-mentioned solutions. One threat is that the hardness of the mathematical problems underlying ECC and RSA cryptosystems will be broken in a relatively near future with the advent of the quantum computers [2]. The other already existing threat is that since IoT devices are susceptible to attacks, their SRAM PUF responses can be obtained and, hence, the device identities can be spoofed even by non-invasive attacks without accessing the device [3–5].

Post-quantum cryptographic algorithms have been proposed to tackle the first threat. The National Institute of Standards and Technology (NIST) is in the process of selecting public-key cryptographic algorithms through a public contest [6]. After three rounds of evaluations that started at the end of 2016, the algorithm that has been recently selected to be standardized for public-key encryption and key establishment is Crystals-Kyber, which will be referred to herein as Kyber.

Behavioral and Physical Unclonable Functions (BPUFs) have been proposed to tackle the second threat [4]. A BPUF can be seen as a multimodal unclonable function, since it generates two or more reproducible, unique and unpredictable responses to given challenges, exploiting the variations produced during the manufacturing process. BPUFs evaluate more distinctive features than conventional PUFs. In addition, the distinctive features evaluated are not only physical but also behavioral. In general, behavioral are more difficult to attack than physical features because they consider more dynamic behaviors. Hence, evaluating behavioral conditions adds security to physical ones. To the best of our knowledge, BPUFs have not been attacked at the moment we write this paper.

This paper proposes a post-quantum protocol for the establishment of a secure communication channel between IoT devices and a central supervisory node (e.g. a server) based on Kyber and SRAM BPUFs. The paper is structured as follows. The related work is presented in Sect. 2. Preliminary concepts regarding the building blocks used in the proposal are shown in Sect. 3 and the proposed protocol is specified in Sect. 4. Experimental results taken with a WiPy board acting as an IoT device are discussed in Sect. 5 and, finally, Sect. 6 concludes the paper.

2 Related Work

In the literature, several works study the employment of post-quantum cryptosystems for IoT applications. Many of them are focused on the importance of lightweight implementations to satisfy limited computation, memory and power resources of IoT devices. As indicated in [7], several works propose accelerators for post-quantum algorithms in IoT devices. In [8], several KEM (Key Encapsulation Mechanism) finalists of the NIST post-quantum cryptography standardization process were implemented in an IoT device. Other works also integrated post-quantum KEM into existing protocols and infrastructures, such as TLS-based [9] or the industrial protocol Open Platform Communications Unified Architecture (OPC UA) in [10].

The establishment of a secure channel to exchange information in a secure way mainly focuses on using shared keys, but the authentication of the parties (IoT devices and gateways or servers) is not so extended. In [10], the mutual authentication between a client and a server was based on pairs of private and public keys whose certificates are compliant to the X.509 standard. In [11], a variation of the McEliece public-key cryptosystem was proposed to be used in an authentication protocol. The authentication protocol was based on the exchange, in a protected way, of the IoT device identifier and nonces generated by the IoT device and the server.

The Kyber KEM algorithm selected by NIST for standardization can be used in authenticated key exchanges as proposed in [12] to provide security against passive adversaries and, particularly, man-in-the-middle attacks. Two authenticated key exchange protocols were proposed: Kyber.UAKE (Unilateral Authenticated Key Exchange), in which one of the parties knows the static (long-term) public key of the other party, and Kyber.AKE (Authenticated Key Exchange), in which each party knows the static (long-term) public key of the other party. The shared key derived at the end of these protocols depends not only on the static (long-term) public keys (and ciphertexts associated), but also on ephemeral public keys (and ciphertexts associated). In this work, we propose the use of the Kyber.UAKE Key Exchange protocol between an IoT device and another entity, such as a remote server, in general.

As commented in Sect. 1, the identity of a device represented by a unique private key can be obtained from a root key reconstructed with a PUF [13, 14]. If a device contains a genuine PUF, it will reconstruct the root key. If the device is counterfeit, it will not be able to recover the key. However, PUFs can be attacked even by non-invasive attacks without accessing the device. Modeling attacks have been reported that replicate the PUF challenges-responses [15]. There are solutions that avoid intercepting the challenge-response pairs by applying encryption or challenge splitting [3]. A non-invasive attack to SRAM PUFs is presented in [5]. The attacker does not need to have direct access to the device. The main vulnerability that they exploit is that physical responses from two SRAM memory chips can have highly correlated properties if they have the same part number. Authors in [5] argue that this can be mainly because of the similarity in its architecture and layout. Also, this correlation can emerge from systematic process variations, especially intra-wafer process variations, which may have deterministic patterns and can be easily to model. Acceptance of errors in SRAM PUF physical responses is cited as another vulnerability. Furthermore, in [16] an SRAM PUF was physically cloned.

In this work, we consider the addition of another security layer by using Behavioral and Physical Unclonable Functions [4]. BPUFs provide two responses, one physical, as conventional PUFs, and another behavioral to the same challenge. On the one hand, if only the physical response is employed, attacks are possible as commented in the previous paragraph. On the other hand, to the best of our knowledge, behavioral responses have not been attacked yet. Note that bit instability is not a desired property in any PUF as it can lower its security, but instead of circumventing it using complex and expensive changes in the SRAM memory as in [5], SRAM BPUFs take advantage of it.

3 Preliminaries

3.1 Kyber Authenticated Key Exchange

Kyber [12] is a public-key encryption scheme defined by key generation, encryption, and decryption algorithms. The key-generation algorithm, Kyber.kGen(), returns a pair of a public and a secret key. The encryption algorithm, Kyber.enc(*pk*, *m*), takes a public key *pk* and a message $m \in \mathcal{M}$ (with $\mathcal{M} \in \{0, 1\}^{256}$) to produce a ciphertext *c*. The decryption algorithm, Kyber.dec(*sk*, *c*), takes a secret key *sk* and a ciphertext *c*, and outputs either a message $m \in \mathcal{M}$ or a special symbol $\perp$ to indicate rejection.

The hard problem underlying the security of Kyber is the Module Learning With Errors (M-LWE), a lattice problem that consists in distinguishing uniform samples $(a_i, b_i) \leftarrow R_q^k \times R_q$ from samples (a_i, b_i) where $a_i \leftarrow R_q^k$ is uniform and $b_i = \mathrm{a}_i^T \cdot \mathrm{s} + e_i$ with $\mathrm{s} \leftarrow \beta_\eta^k$ common to all samples and $e_i \leftarrow \beta_\eta$ fresh for every sample. a_i^T is the transpose of a_i. β_η is a centered binomial distribution with parameter η, where η is even, and the samples are in the interval $[-\eta/2, \eta/2)$. R_q denotes the ring $\mathbb{Z}_q[X]/(X^{256} + 1)$, where $\mathbb{Z}_q$ denotes the ring of integers modulo an integer q (selected in the NIST Kyber proposal as 3329).

A key exchange protocol is a direct application of a Key Encapsulation Mechanism (KEM) because a KEM is a protocol that enables two parties, A and B, to share a session key K. A KEM consists basically of three algorithms: Kyber.kGen, Kyber.encaps, and Kyber.decaps. With Kyber.kGen, the party that initiates the communication (let us assume, it is A) generates an ephemeral public key *pk* for encapsulation and an ephemeral secret key *sk* for decapsulation, and sends *pk* to the other party, B. With Kyber.encaps(*pk*), the party B generates a ciphertext *c* and a session key K by using *pk*. The ciphertext cannot be chosen by B. With Kyber.decaps(*c*, *sk*), the party A decrypts the ciphertext with its private key and obtains the same session key K. Table 1 describes the Kyber.encaps and Kyber.decaps algorithms. F and H are hash functions and the notation $m \leftarrow \{0, 1\}^{256}$ means a 256-bit vector m with uniform distribution. This basic protocol should be improved with authentication to be secure against man-in-the-middle attacks.

Table 1. Steps to performs encapsulation and decapsulation algorithms of Kyber.

Kyber.encaps(pk)	Kyber.decaps(c, sk)
1. $m \leftarrow \{0,1\}^{256}$	1. $m' := \mathrm{Kyber.dec}(c, sk)$
2. $(k, r) := \mathrm{F}(\mathrm{H}(pk), m)$	2. $(k', r') := \mathrm{F}(\mathrm{H}(pk), m')$
3. $c := \mathrm{Kyber.enc}(pk, m; r)$	3. $c' := \mathrm{Kyber.enc}(pk, m'; r')$
4. $K := \mathrm{H}(k, \mathrm{H}(c))$	4. **if** $c' = c$ **then return** $K := \mathrm{H}(k', \mathrm{H}(c))$
5. **return** (c, K)	**else return** $K :=$ random

The Unilateral Authenticated Key Exchange protocol Kyber.UAKE proposed in [12] authenticates one of the parties by its static pair of public and private keys. Figure 1 illustrates this protocol for the case of a non-authenticated IoT Device and an authenticated Remote Server with static pair of public and private keys PK_{RS} and SK_{RS}. Note that the three algorithms Kyber.KGen, Kyber.encaps and Kyber.decaps operations are employed.

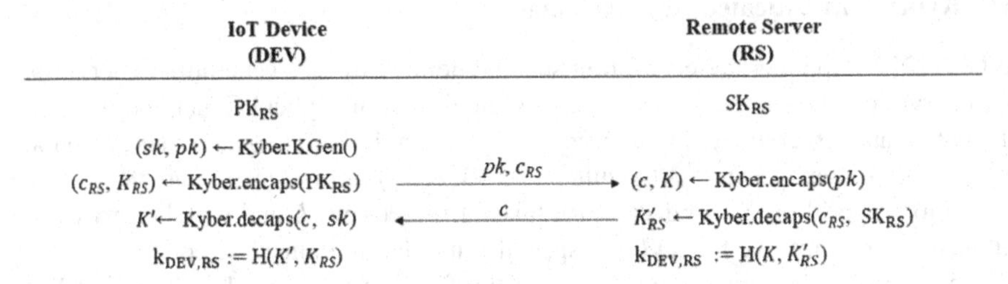

Fig. 1. Kyber Unilateral Authenticated Key Exchange Protocol (Kyber.UAKE).

3.2 Security Increase with SRAM BPUFs

The most extended use of conventional SRAM PUFs is to recover a previously obfuscated secret. In this way, the Kyber.UAKE shown in Fig. 1 could be directly transformed in a Kyber.AKE (with both parties authenticated by long-term public keys) which employed a conventional SRAM PUF by the IoT device to recover its long-term private key. However, we will increase the security of this authentication by using SRAM BPUFs. SRAM BPUFs are more secure because there are two independent ways to check their authenticity, as described in the following.

The b-th bit of the BPUF physical response at measurement i is computed as follows:

$$u_i[b] = \begin{cases} 1 \text{ if BPUF unit } b \text{ meets a physical condition} \\ 0 \qquad\qquad\qquad \text{otherwise} \end{cases} \tag{1}$$

In the case of SRAM BPUFs, the unit b is the memory cell providing the start-up value of the bit b in the BPUF physical response. The physical condition evaluated is if one or the other of the two inverters wins at power up so as to impose a logic 1 or 0.

The b-th bit of the BPUF behavioral response at measurement i is computed as follows:

$$\vartheta_i[b] = \begin{cases} 1 \text{ if } \sum_{j=1}^{R} (u_0[b] \oplus u_j[b]) > 0 \\ 0 \qquad \text{otherwise} \end{cases} \tag{2}$$

In the case of SRAM BPUFs, after taking R sequential start-up measurements, the bit b in the behavioral response is 1 if the start-up value of the memory cell shows bit flipping and it is 0 if it is stable.

The Hamming distance (number of errors) between physical responses of the same SRAM BPUF is small. Hence, it is usual to select an error-correcting code able to correct those errors, and a helper data algorithm to protect the physical response u_0 at the enrollment phase. At enrollment, the following helper data are generated $hD = u_0 \oplus P$, where P results from encoding a secret with the selected error-correcting code ($P = encoding(secret)$). An unbiased physical response has to be selected so as to ensure that the helper data do not leak any information about the secret and can be stored without protection. At verification phase, the fresh physical response, u_i, is combined with the helper data, and the decoder of the error-correcting code is used so that: $secret' = decoding(u_i \oplus hD = u_i \oplus u_0 \oplus P)$. Only if the physical responses come from the same SRAM BPUF, the secret is recovered correctly and also u_0. This is a first way to check if the SRAM BPUF is authentic.

The Hamming distance (HD) between behavioral responses of the same SRAM BPUF is not small. Therefore, no efficient error-correcting codes are able to correct those errors. The Jaccard distance (JD) is better than the Hamming distance to distinguish between genuine and false behavioral responses. The Jaccard distance can be computed from the Hamming distance and the Hamming weights (HW) as follows:

$$JD(\vartheta_0, \vartheta_i) = \frac{2 \cdot HD(\vartheta_0, \vartheta_i)}{HW(\vartheta_0) + HW(\vartheta_i) + HD(\vartheta_0, \vartheta_i)} \quad (3)$$

The Hamming weights of behavioral responses are similar. Hence, we will approximate them by a constant G. If we call $S[b] = \vartheta_i[b] \oplus \vartheta_0[b]$, and N is the number of bits, then we will compute:

$$JD_{app}(S, G) = \frac{2 \cdot \sum_{b=1}^{N} S[b]}{2 \cdot G + \sum_{b=1}^{N} S[b]} \quad (4)$$

Only if the behavioral responses come from the same SRAM BPUF, the approximated Jaccard distance between them is equal or smaller than a threshold JD_{max}. This is the second way to check if the SRAM BPUF is authentic.

4 Proposed Authentication Protocol

4.1 Assumptions

We assume that two parties want to establish a secure communication channel: an IoT Device (DEV) and a Remote Server (RS). The hardware and software of the IoT Device are assumed to be tamper-resistant, because an attacker cannot read the start-up values of the SRAM of the IoT device, and tamper-evident. However, since the IoT Device can be attacked, the BPUF physical response can be cloned. In contrast, the BPUF behavioral response is assumed to better resist attacks because it is defined by more dynamic behaviors.

In this work, we propose the hypothesis that SRAM BPUFs presented in Sect. 2 are 1) *harder to predict*, given that reliability of an SRAM cell is affected by random process variations, and 2) *harder to clone physically*, since bit stability is a property difficult to control.

Also, it is assumed that the IoT Device has a true random number generator, TRNG. This assumption is necessary because in Kyber algorithms there are steps that comprehend the proper generation of random seeds. Also, it is useful for secure generation of internal keys. This TRNG can be included into the IoT Device, or suitable unstable SRAM cells can be used for this purpose as explained in [14]. Since Kyber is considered by the authors as the most suitable post-quantum encryption and key agreement scheme for IoT devices, it is assumed to be used by the IoT Device and the Remote Server. The Remote Server has the secret and public keys, SK_{RS} and PK_{RS}, respectively, which were generated by the key generation algorithm of Kyber. The Remote Server is authenticated as the only one in possession of SK_{RS}. It is assumed that the IoT Device has already the public key PK_{RS} before any authentication phase is performed. The challenge for the BPUF of the IoT Device is assumed to be set. The Remote Server has access to enrollment data of the IoT Device, and it has the threshold Jaccard distance JD_{max}.

Attacks can be done against the IoT Device and the communication channel, but the Remote Server is assumed to be secure. The Remote Server is assumed to be honest, but can be curious. For this reason, it is convenient not to store any BPUF response directly in the Remote Server.

4.2 Enrollment Phase

The enrollment phase is done in a controlled environment and, hence, free of attacks. This phase contains two major steps, to generate the data needed for authenticating the IoT Device through the BPUF physical response and to generate the data needed for authenticating the IoT Device through the BPUF behavioral response.

The Remote Server generates an identifier ID_{DEV} for the IoT Device and sends it to it together with the enrollment request $RQST$. This identifier is used only for indexing purposes. The device obtains the BPUF physical response u_0 with the function BPUF.getPhysicalResponse. As another authentication factor, the IoT Device obtains the BPUF behavioral response ϑ_0 using the function BPUF.getBehavioralResponse. Also, the device generates a secret key k_{DEV} only known by it with its TRNG, using the function TRNG.genRandomSeed. The final non-sensitive data representing the identity of the device, ID^0_{BPUF}, results from XOR-ing the behavioral response ϑ_0 with a key K_{DEV} obtained from k_{DEV} and the index ID_{DEV}, using the pseudo-random function PRF. The length of the output of this function, K_{DEV}, should be the same as the length of ID^0_{BPUF}. The secure IoT Device identifier, ID^0_{BPUF}, is irreversible in the sense that neither the physical nor the behavioral responses can be obtained from it. It shows unlinkability in the sense that cannot be known the BPUF or device that generated it. Besides, it is different for different verifiers. Finally, it shows revocability because it can be revoked by changing the random secure identifier ID^0_{BPUF} in a new enrollment process. The helper data hD is generated by obfuscating k_{DEV}, using the physical response u_0, with the function BPUF.helperData. It is noted that hD are not sensitive data and can be computed by using an Error Correcting Code. The Hamming weight of ϑ_0, G, is calculated by using the function HammingWeight. The device stores ID_{DEV} and hD. They can be stored in a common non-volatile memory because they are not sensitive data. The Remote Server stores G, ID^0_{BPUF}, and ID_{DEV} as an index, and it has already JD_{max} as the authentication threshold. The data ID^0_{BPUF} are stored in a secure way since they represent the protected identity of the device. If the Remote Server is curious, it cannot obtain any information from these data. Even if those data are leaked from the Remote Server, no information can be obtained, neither the devices linked to them, and the system is restored by revoking the identifiers. Figure 2 illustrates the enrollment phase.

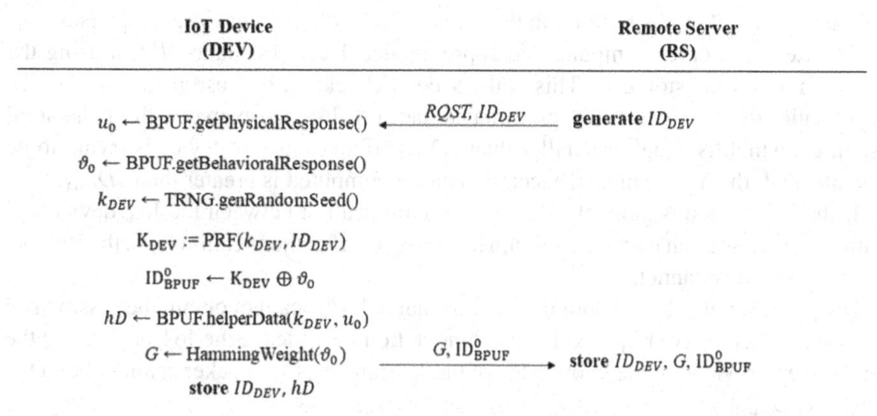

Fig. 2. Proposed enrollment phase between the IoT Device and the Remote Server.

4.3 Authentication Phase

The authentication phase is done every time the Remote Server wants to establish a secure communication with the IoT Device. The IoT Device starts the Unilateral Authenticated Key Exchange protocol Kyber.UAKE to authenticate the Remote Server and establish a session key $k_{DEV,RS}$ with it. Note that there are two interactions in the UAKE phase as was shown in Fig. 1. In addition to pk and c_{RS}, the IoT Device sends its index ID_{DEV} to the Remote Server in the first transmission of the UAKE.

The Remote Server first sends a nonce seed n_{RS}'(generated by the function generateNonce) to the IoT Device to guarantee freshness. The nonce seed is encrypted with the session key $k_{DEV,RS}$. This nonce seed will be expanded later to the length of the ID^i_{BPUF} using a pseudo-random function and the index ID_{DEV} resulting the nonce n_{RS}. The IoT Device obtains a fresh BPUF physical response u_i and employs it and the stored helper data hD to recover the response k_{DEV} generated at enrollment phase, using the decoder of the employed Error Correcting Code and the Helper Data Algorithm. Then, with the stored index ID_{DEV}, it reconstructs the key K_{DEV}. If this step, executed using the function BPUF.reconstructKey, is done by a genuine device, the key reconstructed is the same key employed at enrollment.

Also, the IoT Device obtains a fresh BPUF behavioral response ϑ_i and obfuscates it not only with the key K_{DEV} but also with the nonce n_{RS}. The device encrypts the identifier result, ID^i_{BPUF}, with the session key $k_{DEV,RS}$, and sends that to the Remote Server. Thus, the exchanged information is confidential.

The Remote Server computes the value S with the stored $\mathrm{ID}_{\mathrm{BPUF}}^{0}$ that is retrieved using the index ID_{DEV} with the function retrieveDataID. This is the first checking, because if the fresh physical BPUF response is authentic, the value S verifies that:

$$S = ID_{BPUF}^{i} \oplus ID_{BPUF}^{0} \oplus n_{RS} = \vartheta_{x}^{i} \oplus \vartheta_{x}^{0} \tag{5}$$

If an impostor device is trying to be authenticated with an erroneous physical response, the result S has nothing to do with the XOR of the behavioral responses.

The Remote Server computes the approximated Jaccard distance JD_{app} using the computed S and the stored G. This is the second checking, because if the IoT Device is authentic, the result is the approximated Jaccard distance between the behavioral responses, which is equal or smaller than JD_{max}. If an impostor device is trying to be authenticated, the approximated Jaccard distance computed is greater than JD_{max}.

If the IoT device is authenticated, the communication between the IoT device and Remote Server goes on using the session key $k_{DEV,RS}$. If authentication fails, the Remote Server closes the channel.

If an attacker attacks the communication channel, s/he cannot obtain the session key because the secret keys required for its computation never leave the IoT device and the Remote Server. Without the knowledge of the session key, the attacker cannot obtain the identifier $\mathrm{ID}_{\mathrm{BPUF}}^{i}$.

The steps of the authentication phase are summarized in Fig. 3.

IoT Device (DEV)		**Remote Server (RS)**
PK_{RS}, ID_{DEV}, hD		$SK_{RS}, JD_{max}, ID_{DEV}, G, ID_{BPUF}^{0}$
$(sk, pk) \leftarrow$ Kyber.KGen()		
$(c_{RS}, K_{RS}) \leftarrow$ Kyber.encaps(PK_{RS})	ID_{DEV}, pk, c_{RS} $\rightarrow$	$(c, K) \leftarrow$ Kyber.encaps(pk)
$K' \leftarrow$ Kyber.decaps(c, sk)	$\leftarrow$ c	$K'_{RS} \leftarrow$ Kyber.decaps(c_{RS}, SK_{RS})
$k_{DEV,RS} := H(K', K_{RS})$		$k_{DEV,RS} := H(K, K'_{RS})$
$u_i \leftarrow$ BPUF.getPhysicalResponse()	$\leftarrow$ $Enc(k_{DEV,RS}, n'_{RS})$	$n'_{RS} \leftarrow$ generateNonce()
$\vartheta_i \leftarrow$ BPUF.getBehavioralResponse()		$n_{RS} \leftarrow PRF(n'_{RS}, ID_{DEV})$
$k'_{DEV} :=$ BPUF.reconstructKey(u_i, hD)		$ID_{BPUF}^{0} \leftarrow$ retrieveDataID(ID_{DEV})
$K'_{DEV} := PRF(k'_{DEV}, ID_{DEV})$		
$n_{RS} \leftarrow PRF(n'_{RS}, ID_{DEV})$		
$ID_{BPUF}^{i} \leftarrow K'_{DEV} \oplus \vartheta_i \oplus n_{RS}$	$Enc(k_{DEV,RS}, ID_{BPUF}^{i})$ $\rightarrow$	$S \leftarrow ID_{BPUF}^{i} \oplus ID_{BPUF}^{0} \oplus n_{RS}$
		compute JD_{APP} from S and G
		if $JD_{APP} > JD_{max}$: failed communication

Fig. 3. Proposed authentication phase between the IoT Device and the Remote Server.

5 Discussion and Experimental Results

Table 2 shows our proposal compared the recent related work commented in Sect. 2. Clearly, our proposal is the only one that considers hardware security along with post-quantum security.

Table 2. Comparison with the recent related work cited in Sect. 2.

	[7]	[8]	[9]	[10]	[11]	Ours
Quantum-resistant	Yes	Yes	Yes	Yes	Yes	Yes
Standardized	Yes	Yes	Yes	Yes	No	Yes
Type of cryptography	Lattice-based	Lattice-based	Lattice-, code-, isogeny-based	Lattice-based	Code-based	Lattice-based
Security level	128-192-256	128	128-192-256	128-192-256	80-128	128-192-256
Hardware security	No	No	No	No	No	Yes
Mutual authentication	No	No	Yes	No	Yes	Yes

An initial evaluation of the proposal was carried out in an Espressif's ESP32 microcontroller integrated in a Pycom's WiPy 3.0 board. This evaluation is centered in the IoT device. The ESP32 microcontroller is widely used in IoT applications due to its low cost and its integrated WiFi and Bluetooth communication modules. The microcontroller has 512 KB of internal SRAM and concerning non-volatile storage, it is included a Flash memory of 8 MB in the board. It was chosen a clock frequency of 160 MHz for taking execution time measurements. Kyber KEM was implemented using the reference code found in [17] provided by its authors. Kyber KEM has three possible security levels: II, III and IV, associated to 128, 192 and 256 bits. A region of the internal SRAM of the ESP32 was used as BPUF and three different boards were used for an initial characterization.

BPUF responses of 8,192 bits were evaluated. Regarding the physical response of the BPUF, it presents an average intra fractional Hamming distance of 0.0324. Hence, a repetition error correcting code of 32 bits was selected. The steps for this selection can be found in [4]. The behavioral response has an average inter Jaccard distance of 0.8997 and an average intra Jaccard distance of 0.3740. The corresponding threshold value JD_{max} selected is 0.5581.

Concerning execution times, the authentication phase is the most important since enrollment phase is done only once while the authentication is done several times in an application environment. It was found that operations related with Kyber KEM were the costliest. They are shown in Table 3. The execution time of other operations are very low in comparison.

Concerning the communication bandwidth, the biggest data to be sent are those related with Kyber UAKE and ID^{i}_{BPUF}. This information is summarized in Table 4. Since a key of 256 bits is obfuscated with the physical part of the BPUF, a Helper Data of 1,024 bytes needs to be stored in the non-volatile memory of the IoT device.

Table 3. Execution times in ms of operations performed by the IoT Device.

	II	III	IV
KGen + encaps	18.30	29.56	44.70
decaps + hash	11.99	18.47	26.95

Table 4. Sizes in bytes of data that affect communication bandwidth.

	II	III	IV
$pk + \mathrm{CRS}$	1,568	2,272	3,136
	II	III	IV
c	768	1,088	1,568
n'_{RS}		32	
ID_{DEV}		32	
$\mathrm{ID}_{\mathrm{BPUF}}$		1,024	

6 Conclusions

This work proposes a protocol between an IoT Device and a Remote Server that combines the Kyber Unilateral Authenticated Key Exchange (Kyber.UAKE) protocol with BPUFs. The IoT Device is authenticated by SRAM BPUFs. The security of the communication is increased because the authenticity of the IoT Device is proved not only by a physical, but also, by a behavioral way. The protocol was implemented by considering WiPy boards as IoT devices. BPUFs responses were evaluated in nominal conditions to select the error-correcting code and the authentication threshold. Execution times and communication bandwidths were obtained for BPUFs and Kyber constructions with security levels of 128, 192 and 256 bits. These results proved the feasibility of the proposal in common IoT devices. Thanks to the inclusion of BPUFs, the authentication is much stronger.

Acknowledgements. This research was conducted thanks to Grant PDC2021-121589-I00 funded by MCIN/AEI/https://doi.org/10.13039/501100011033 and the "European Union NextGenerationEU/PRTR", and Grant PID2020-119397RB-I00 funded by MCIN/AEI/ https://doi.org/10.13039/501100011033. The work of Roberto Román was supported by VI Plan Propio de Investigación y Transferencia through the University of Seville.

References

1. Mall, P., Amin, R., Das, A.K., Leung, M.T., Choo, K.-K.R.: PUF-based authentication and key agreement protocols for IoT, WSNs, and Smart Grids: a comprehensive survey. IEEE Internet Things J. **9**(11), 8205–8228 (2022)
2. Buchmann, J., Ding, J. (eds.): PQCrypto 2008. LNCS, vol. 5299. Springer, Heidelberg (2008). https://doi.org/10.1007/978-3-540-88403-3
3. Ebrahimabadi, M., Younis, M., Karimi, N.: A PUF-based modeling-attack resilient authentication protocol for IoT devices. IEEE Internet Things J. **9**(5), 3684–3703 (2022)

4. Prada-Delgado, M.A., Baturone, I.: Behavioral and Physical Unclonable Functions (BPUFs): SRAM example. IEEE Access **9**, 23751–23763 (2021)
5. Bahar Talukder, B.M.S., Ferdaus, F., Rahman, M.T.: Memory-based PUFs are vulnerable as well: a non-invasive attack against SRAM PUFs. IEEE Trans. Inf. Forensics Secur. **16**, 4035–4049 (2021)
6. NIST CSRC, Post-Quantum Cryptography. https://csrc.nist.gov/projects/post-quantum-cryptography. Accessed 28 July 2022
7. Seyhan, K., Nguyen, T.N., Akleylek, S., Cengiz, K.: Lattice-based cryptosystems for the security of resource-constrained IoT devices in post-quantum world: a survey. Clust. Comput. 1–20 (2021). https://doi.org/10.1007/s10586-021-03380-7
8. Septien-Hernandez, J.-A., Arellano-Vazquez, M., Contreras-Cruz, M.A., Ramirez-Paredes, J.-P.: A Comparative study of post-quantum cryptosystems for Internet-of-Things applications. Sensors **22**(2), 489 (2022)
9. Schöffel, M., Lauer, F., Rheinländer, C.C., When, N.: Secure IoT in the era of quantum computers—where are the bottlenecks? Sensors **22**(7), 2484 (2022)
10. Paul, S., Scheible, P.: Towards post-quantum security for cyber-physical systems: integrating PQC into industrial M2M communication. In: Chen, L., Li, N., Liang, K., Schneider, S. (eds.) ESORICS 2020. LNCS, vol. 12309, pp. 295–316. Springer, Cham (2020). https://doi.org/10.1007/978-3-030-59013-0_15
11. Chikouche, N., Cayrel, P.-L., Mboup, E.H.M., Boidje, B.O.: A privacy-preserving code-based authentication protocol for Internet of Things. J. Supercomput. **75**(12), 8231–8261 (2019). https://doi.org/10.1007/s11227-019-03003-4
12. Bos, J., et al.: CRYSTALS – Kyber: a CCA-secure module-lattice-based KEM. In: 2018 IEEE European Symposium on Security and Privacy (EuroS&P), pp. 353–367. IEEE (2018)
13. Cambou, B., et al.: Post quantum cryptographic keys generated with physical unclonable functions. Appl. Sci. **11**(6), 2801 (2021)
14. Román, R., Arjona, R., Arcenegui, J., Baturone, I.: Hardware Security for eXtended Merkle Signature Scheme using SRAM-based PUFs and TRNGs. In: 2020 32nd International Conference on Microelectronics (ICM), pp. 1–4. IEEE (2020)
15. Zerrouki, F., Ouchani, S., Bouarfa, H.: A survey on silicon PUFs. J. Syst. Architect. **127**, 102514 (2022)
16. Helfmeier, C., Boit, C., Nedospasov, D., Seifert, J. -P.: Cloning physically unclonable functions. In: 2013 IEEE International Symposium on Hardware-Oriented Security and Trust (HOST), pp. 1–6. IEEE (2013)
17. GitHub, pq-crystals/Kyber. https://github.com/pq-crystals/kyber. Accessed 28 July 2022

Effective Segmentation of RSSI Timeseries Produced by Stationary IoT Nodes: Comparative Study

Pooria Madani[1(✉)] and Natalija Vlajic[2]

[1] Faculty of Business and IT, University of Ontario Institute of Technology, Oshawa, Canada
pooria.madani@ontariotechu.ca

[2] Department of Electrical Engineering and Computer Science, York University, Toronto, Canada
vlajic@yorku.ca

Abstract. The Received Signal Strength Indicator (RSSI) timeseries have been used as a primary variable in many cybersecurity applications, such as wireless-node profiling for the purpose of authentication, localization, and physical security perimeter monitoring. Previous research on the use of RSSI-based wireless node profiling assumes that RSSI timeseries are stationary and independent identically distributed (i.i.d.). Unfortunately, in real-world environments, this assumption is far from the truth and would negatively impact the performance of any system or application built on idealized models of RSSI timeseries data. In other words, a set of real-world RSSI values (depending on the variability of noise produced by objects in the environment) are typically made of sub-segments each with its own statistical characteristics (e.g., mean and variance). Therefore, before any modelling attempt, one must consider breaking down a given RSSI dataset into its constituting sub-segments. Unfortunately, the effect of environmental variables on RSSI values tend to be random, which makes the problem of RSSI timeseries segmentation even more challenging. Thus, it is necessary to study the effectiveness of existing notable timeseries segmentation algorithms against a dataset of RSSI values. The main contributions of our work are that (1) we have demonstrated the non-stationary nature of RSSI timeseries by collecting samples from a real-world IoT network, and (2) through real-world experimentation we have compared the effectiveness of notable timeseries segmentation methods for the discovery of sub-segments in a RSSI timeseries dataset. Our work highlights the importance of accurate detection of change points in RSSI timeseries, which can further facilitate optimal selection and performance of the respective system's cost and objective functions. Finally, we demonstrate that the ℓ_1 cost function can capture a meaningful relationship between neighboring data points in a RSSI timeseries and can result in a stable segmentation across different search methods.

Keywords: Timeseries · Segmentation · RSSI · Preprocessing

© The Author(s), under exclusive license to Springer Nature Switzerland AG 2022
W. Li et al. (Eds.): ADIoT 2022, LNCS 13745, pp. 84–101, 2022.
https://doi.org/10.1007/978-3-031-21311-3_4

1 Introduction

The received signal strength indicator (RSSI) is a numerical variable that is directly influenced by the transmission power and the location of the transmitter as well as different environmental variables such as obstacles situated between the transmitter and the receiver. As suggested in a number of earlier research works (e.g., [1–3]), RSSI values can be used to create the fingerprint profile of each device in a wireless network and subsequently use the developed profiles in many cybersecurity applications such as node authentication [4,5], COVID-19 tracking [6], and localization [7]. Most of the wireless node profiling studies utilizing RSSI implicitly assume that: (1) RSSI samples received from a non-moving transmitting device form a stationary time-series with normally distributed variance, and (2) RSSI values are independent and identically distributed (i.i.d.) samples from an unknown normal distribution. However, in our recently conducted study [5], the two assumptions (RSSI samples are stationary and i.i.d.) have come under scrutiny. Namely, through our extensive real-world experimentation, we have observed that RSSI values measured by a receiving node are highly affected by changes (e.g., moving objects) in their operating environments. In particular, we have observed that people moving around (or there being no moving objects in a house) have a noticeable effect on RSSI values of IoT devices deployed in a residential environment. As a result, the variance of the RSSI time-series changes significantly when occupants are present and move around the property (refer to Fig. 1 where there are two different segments, one with lower volatility than the other). Furthermore, it is clear from Fig. 1 that there is a correlation between neighboring RSSI values; therefore, it would be hard to justify the claim that neighboring RSSI values are independent.

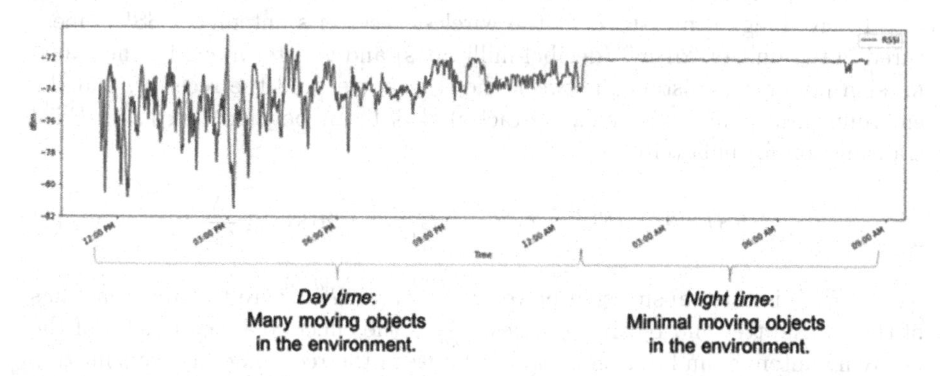

Fig. 1. RSSI values of an IoT device deployed in a residential property with routine movements of occupants in a 24 h period.

Except in a few use cases where there are no moving objects in the environment (e.g., farmland monitoring), most real-world IoT networks deploy computing/sensing nodes in environments with some number of movable objects. Thus, in order to account for changes in RSSI values as shown in Fig. 1, it is necessary to perform timeseries segmentation before using RSSI values for profiling and modelling purposes.

Fundamentally, RSSI timeseries are substantially different from well-known structured timeseries such as those studied in the field of medicine or finance. The effect of environmental variables on RSSI values is seemingly random, which can make the problem of segmentation comparatively difficult. As a result, we find it necessary to study the effectiveness of existing notable timeseries segmentation approaches against a dataset of RSSI values. As our main contribution, we have surveyed and studied the effectiveness of notable timeseries segmentation algorithms on RSSI timeseries of typical IoT nodes based on a real-world dataset.

The content of this paper is organized as follows: in Sect. 2, we define the problem statement and survey notable cost functions and search methods that are predominantly used in timeseries segmentation. Moreover, we introduce the evaluation criteria that we are going to use to compare the effectiveness of segmentation schemes when applied to our RSSI timeseries dataset. In Sect. 3, we define the test environment that we have used to generate and collect the RSSI samples and present the comparison results of surveyed segmentation schemes from Sect. 2. Finally, in Sect. 4 we go over the potential future direction of this work.

2 Survey of Change Point Detection Algorithms

2.1 Problem Statement

The received signal strength indicator (RSSI) is a measurement of the power of a received signal measured at the wireless receiver's antenna. RSSI is measured in the unit of "dBm" (decibel-milliwatts) and is determined by the transmission power, the distance between the transmitter and the receiver, and the environmental conditions (e.g., obstacles). RSSI can be modelled using Friis' transmission formula (cite) as

$$RSSI^{[dB]} = P_t^{[dB]} + G_t^{[dBi]} + G_r^{[dBi]} + 20\log_{10}(\frac{\lambda}{4\pi d}) \qquad (1)$$

where $P_t^{[dB]}$ is the transmission power, $G_t^{[dBi]}$ and $G_r^{[dBi]}$ are the antenna gains of the transmitter and receiving nodes, $\frac{\lambda}{4\pi d}$ is the effective aperture area of the receiving antenna, and d is the distance between the receiving and transmitting antennas.

Let us consider a non-stationary timeseries of RSSI values $y = \{y_1, \ldots, y_T\}$ that has T samples. The signal y is assumed to be piecewise stationary, i.e., that some characteristics of the process change abruptly at some unknown instants $\tau = \{t_1^* < t_2^* < \ldots < t_K^*\}$. Change point detection is the process of estimating the indexes $t_k^* \in \tau$. Depending on the environmental variabilities that could affect RSSI, the number of changes (i.e., $K = |\tau|$) may or may not be known, in which case it has to be estimated.

A timeseries can change its statistical properties (e.g., mean and/or variance) from one period to the next due to some structural changes in the process producing the observed values in the series. For example, RSSI values (i.e., y) can be highly affected by changes in the environmental variables such as the presence/absence of moving objects. For any model-driven application built on the assumption of a stationary time series as an input, successful detection of such structural changes in the given time series before the modelling process is critical. In particular, many cybersecurity applications (e.g., authentication and intrusion detection) rely on the assumption of a stationary RSSI time series, and as a preprocessing step (depicted in Fig. 2), it is important to segment the given timeseries into multiple homogeneous sets by estimating the indexes $t_k^* \in \tau$ before any modelling exercise.

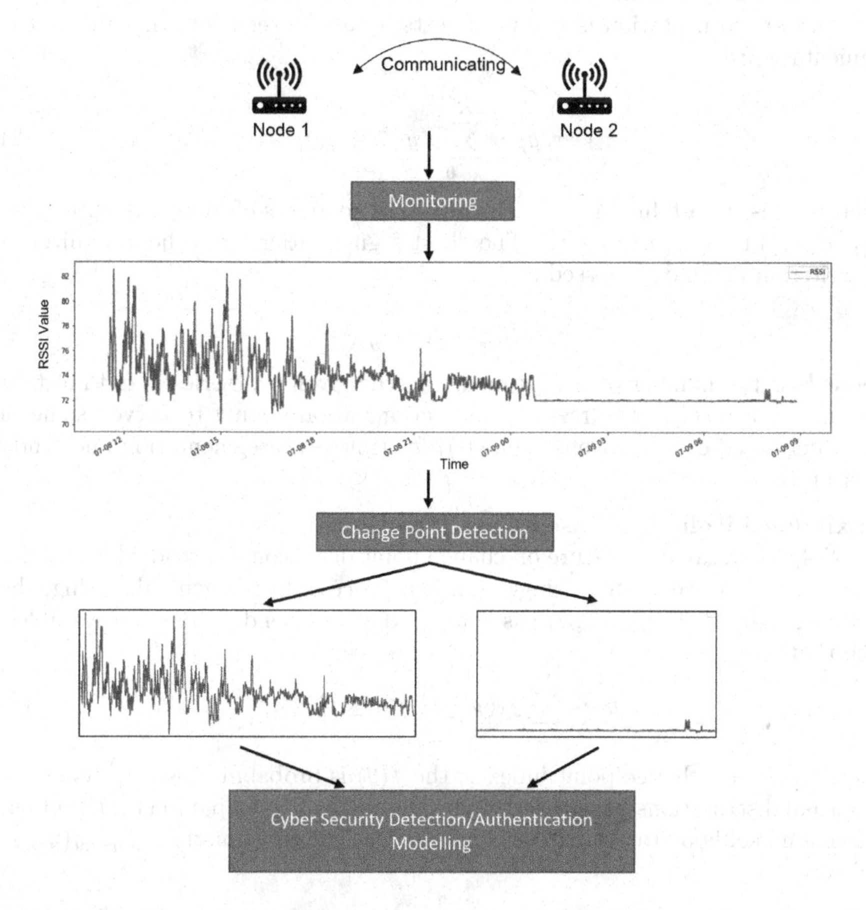

Fig. 2. RSSI values of an IoT device deployed in a residential property with routine movements of occupants in a 24 h period.

Change point detection (i.e., timeseries segmentation) can be formulated as a search problem governed by a cost function that must be minimized. The cost function measures *homogeneity* in the discovered segments within a timeseries and dictates the type of changes that can or should be detected. The search method is the resolution procedure for the discrete optimization problem defined by the cost function. In the rest of this paper, we are going to survey some of the well known cost functions and search methods for *change point detection* in timeseries and study their effectiveness on a real-world RSSI timeseries dataset extracted from an actual IoT network.

2.2 Cost Functions

Choosing the best possible segmentation τ for a given y can be formulated as a model selection problem governed by a quantitative criterion $V(\tau, y)$ that must be minimized. It is widely assumed [8] that the criterion function $V(\tau, y)$ for a particular segmentation is a sum of costs of all the segments that define the segmentation:

$$V(\tau, y) = \sum_{k=0}^{K} c(y_{t_k,\ldots,t_{k+1}}), \tag{2}$$

where $c(\cdot)$ is a cost function which measures goodness-of-fit of the sub-signal $c(y_{t_k,\ldots,t_{k+1}})$ to a specific model. The "best segmentation" τ is the minimizer of the criterion (2) and expressed as

$$min_{|\tau|=K} V(\tau, y), \tag{3}$$

where K is the number of true segments in the given timeseries y and must be known in advance. In the rest of this section, we are going to survey some of the widely used cost functions (i.e., $c(\cdot)$) for timeseries segmentation and study their efficacy in Sect. 3.

Maximum Likelihood-Based Cost Functions

Many of the research literature on change point detection has studied timeseries whose values are generated using a random process. In the general setting, the observed signal $y = \{y_1, \ldots, y_T\}$ is composed of independent random variables, such that

$$y_t \sim \sum_{k=0}^{K^*} f(\theta_k) I(t_k^* < t \leq t_{k+1}^*), \tag{4}$$

where the t_k^* are change point indexes, the $f(\theta)$ is probability density functions of normal distributions parameterized by the vector-valued parameter θ and use maximum likelihood to estimate this parameter. The cost function $c_{normal}(y_{a\ldots b})$ is defined as

$$c_{normal}(y_{a\ldots b}) = -sup_{\theta} \sum_{t=a+1}^{b} \log f(y_t|\theta). \tag{5}$$

The parameter θ in (5) represents the quantity of interest where the value changes abruptly at the unknown t_k^* and it is the job of the change detector algorithm to estimate this parameter (or set of such parameters). Using maximum likelihood estimation, one can estimate parameters for (5) for different types of statistical distributions. In this paper, however, we only consider θ belongs to the family of Gaussian distributions.

The cost function can be rewritten as

$$c_{\ell 2}(y_{a \dots b}) = \sum_{t=a+1}^{b} ||y_t - \bar{y}_{a \dots b}||_2^2, \tag{6}$$

when one considers Gaussian distribution formulation to estimate (5) with fixed variance, where $\bar{y}_{a \dots b}$ is the empirical mean of the sub-series $y_{a \dots b}$. This cost function is heavily used in DNA array data and geology analysis and is sometimes referred to as the quadratic error loss function [9].

Piecewise Linear Regression-Based Cost Functions

One can argue that the piecewise relationship between elements of y can be described by a linear model. Under such assumption, one can also argue that this relationship can change abruptly at some known points in y,

$$\forall t, t_k^* < t \le t_{k+1}^*, \quad y_t = x_t^{'} u + k + z_t^{'} + \epsilon_t \qquad (k = 0, \dots, K^*) \tag{7}$$

where $u_k \in R^p$ and $v_k \in R^q$ are unknown regression parameters (i.e., estimated using linear regression) and ϵ_t is noise. As a result, change point detection can be done by fitting a linear regression on each segment of the signal while optimizing for the following cost function:

$$c_{linear}(y_{a \dots b}) = \sum_{t=a+1}^{b} (y_t - x_t^{'} u, y_t - z_t^{'} v)^2. \tag{8}$$

One could also consider optimizing for the least absolute deviations criterion instead of least square criterion in order to define the following cost function:

$$c_{linear, \ell_1}(y_{a \dots b}) = \sum_{t=a+1}^{b} ||(y_t - x_t^{'} u, y_t - z_t^{'} v)||. \tag{9}$$

Cost function (9) is often used to detect structural changes in economic and financial data with heavy tail noise distributions [8]. One could also define the cost function to capture piecewise autoregressive in y. The autoregression model (denoted as c_{AR}) is a popular representation of random processes, where each element $y_t \in y$ depends linearly on the previous variables $x_t = [y_{t-1}, y_{t-2}, \dots, y_{t-p}]$ governed by a lagged period p.

$$c_{AR}(y_{a \dots b}) = \sum_{t=a+1}^{b} ||y_t - x_t^{'} u||^2 \tag{10}$$

The cost function (10) can detect shifts in the autoregression coefficients of a non-stationary process and is widely used in EEG/ECG and functional magnetic resonance imaging (fMRI) timeseries analysis [10].

Kernel-Based Cost Functions
Kernel-based cost functions rely on non-parametric modelling of y as

$$y_t \sim \sum_{k=0}^{K^*} F_k I(t_k^* < t \leq t_{k+1}^*), \tag{11}$$

where the t_k^* are change point indexes and the F_k is a cumulative distribution function (c.f.d.). Let $k(\cdot,\cdot) : R^d \times R^d \rightarrow R$ be an arbitrary kernel function mapping signal y onto a reproducing Hilbert space $\mathcal{H}$, then a generic kernel-based cost function is defined as

$$c_{kernel}(y_{a...b}) = \sum_{t=a+1}^{b} ||\phi(y_t) - \bar{\mu}_{a...b}||^2_{\mathcal{H}} \tag{12}$$

$$\phi(y_t) = k(y_t, \cdot) \in \mathcal{H}, \tag{13}$$

where $\bar{\mu}_{a...b} \in \mathcal{H}$ is the empirical mean of the embedded signal $\{\phi(y_t)\}_{t=a+1}^{b}$.

In this paper we are going to study the use of two specific kernels $k(\cdot,\cdot)$ for the purpose of change point detection on RSSI time series, as follows:

1. Linear kernel: $k(x, y) = x^T y$ which makes the cost function c_{kernel} equivalent to $c_{\ell 2}$.
2. Gaussian kernel: $k(x, y) = exp(-\gamma||x - y||^2)$ with $x, y \in R^d$ and where $\gamma > 0$ is the bandwidth parameter.

2.3 Search Methods

The second defining element of "change point detection" is the search method. The search methods are in charge of finding the optimal segmentation τ of the given timeseries y by solving the equation (3) under a given cost function. Finding an exact solution to (3) (i.e., the optimal segmentation), depending on the chosen cost function, can be computationally expensive. Therefore, aside from *optimal search* methods, there have been many *approximation-based* search methods have been proposed in the literature [8] that we have surveyed and used in this paper.

Dynamic Programming
Using dynamic programming and additive nature of the objective function $V(\cdot)$ in (3), one can find the optimal solution by recursively solving sub-problems by making the following observation [8]:

$$\min_{|\tau|=K} V(\tau, y = y_{0...T}) = \min_{t \leq T-K} \left[c(y_{0...T}) + \min_{|\tau|=K-1} V(\tau, y_{t...T}) \right]. \tag{14}$$

In other words, (14) assumes that the first change point of the optimal segmentation is easily computed if the optimal partitions with $K-1$ elements of $y_{t\ldots T}$ are known. And, carries out this logic recursively until the end of y.

Window Sliding

The window sliding algorithm [8] can be considered a fast approximation to the dynamic programming algorithm discussed earlier. The main idea behind this approach is that one can compute the discrepancy between two adjacent windows that slide along y. In other words, when two windows cover dissimilar segments, the discrepancy reaches large values - computed using a given cost function $c(\cdot)$:

$$d(y_{a\ldots t}, y_{t\ldots b}) = c(y_{a\ldots b}) - c(y_{a\ldots t}) - c(y_{t\ldots b}). \tag{15}$$

We refer the reader to [8] for details of implementation of window sliding change detection algorithm.

Binary Segmentation

The binary segmentation algorithm is a search procedure that recursively splits y until a stopping criterion is met. Let $\hat{t}^{(1)}$ denote the first change point estimation computed as follows:

$$\hat{t}^{(1)} = \underset{1\leq t<T-1}{\operatorname{argmin}}\; c(y_{0\ldots t}) + c(y_{t\ldots T}). \tag{16}$$

This is a *greedy* search algorithm given that it searches the change point that lowers the sum of the costs at each step; and recursively applies (16) to each split.

Bottom-Up Segmentation

The bottom-up segmentation algorithm is an approximation search procedure for finding the solution to (3). It begins by splitting y into many small subsets (size of subsets are passed as an argument to the algorithm) and sequentially merges them up until there are only K change points. In other words, using (15) as discrepancy measure, at each level, two segments with the lowest score are merged until K change points remain.

Matrix Profile Segmentation

The matrix profile algorithm [11] is an all-pairs similarity finding algorithm in timeseries that uses a novel data structure to store similarity scores between points in a timeseries. As a side effect, one could use the assembled data structure (as a result of executing this algorithm) to segment a timeseries and detect points of change without having any advanced knowledge on the total number of separable segments in the timeseries. Contrary to the previously discussed search algorithms, the matrix profile approach does not optimize for (3), and segmentation of the timeseries is done through counting number of arcs crossing over each cell in the data structure.

The algorithm starts by dividing the timeseries y into n non-overlapping subseries $s = y_{1\ldots n}, y_{n+1\ldots 2n}, \ldots$, and for each $s_i \in s$ one would find another timeseries $s_j \in s$ where $i < j$ and $c(s_i, s_j) < c(s_i, s_k)$ for all values of k other than $k \neq i$ and $k \neq j$; let C to be the tuple-set storing such (i, j) pairs. Then, a change point s_c is any window that one could not find an index pair $(l, m) \in C$ such that $l < c < m$. In other words, s_c is considered change point window if and only if one could not find any window predeceasing s_c that is ruled similar to another window proceeding s_c.

Surprisingly, as will be demonstrated in Sect. 3, change point detection using the matrix profile algorithm is very effective in timeseries segmentation and the computational complexity of the algorithm is linear in respect to the number of the elements in y. Such computational complexity makes this algorithm an excellent candidate for implementation on low power computing nodes (e.g., IoT nodes) for change point detection in very long timeseries.

2.4 Evaluation

In this section we are going to provide an overview of evaluation functions that measure *fitness* of timeseries segmentation approaches compared to the best-known segmentation (i.e., the ground truth). In the rest of the section, let $T^* = \{t_1^*, \ldots, t_{K^*}^*\}$ denotes the set of true change points, and $\hat{T} = \{\hat{t_1}, \ldots, \hat{t}_{\hat{K}}\}$ denotes the set of estimated change points. Note, that the cardinals of each set, K^* and $\hat{K}$, are not necessarily equal.

RandIndex

The RandIndex [12] is a method for measuring accuracy of segmentation of a timeseries by computing the average similarity between the predicted breakpoint set $\hat{T}$ and T^*. Formally, given a breakpoint set T, we define the set of grouped indexes and the set of non-grouped indexes as follows:

$$\begin{aligned} gr(T) =&\{(s,t), 1 \leq s < t \leq T \text{ s.t. } s \text{ and } t \\ &\text{belong to the same segment according to } T\}, \end{aligned} \tag{17}$$

$$\begin{aligned} ngr(T) =&\{(s,t), 1 \leq s < t \leq T \text{ s.t. } s \text{ and } t \\ &\text{belong to the different segment according to } T\}. \end{aligned} \tag{18}$$

Then, RandIndex is defined as follows:

$$RandIndex(\hat{T}, T^*) = \frac{|gr(\hat{T}) \cap gr(T^*)| + |ngr(\hat{T}) \cap ngr(T^*)|}{T(T-1)} \tag{19}$$

RandIndex is normalized between 0 (for total disagreement) and 1 (for total agreement) between $\hat{T}$ and T^*.

Hausdorff

The Hausdorff metric measures greatest temporal distance between a change point and its prediction and is defined as:

$$\text{Hausdorff}(\hat{T}, T^*) = \max\{\max_{\hat{t} \in \hat{T}} \min_{t^* \in T^*} |\hat{t} - t^*|, \max_{t^* \in T^*} \min_{\hat{t} \in \hat{T}} |\hat{t} - t^*|\}. \tag{20}$$

The Hausdorff is considered a metric for measuring robustness of detection [13]. It is large when a change point from $\hat{T}$ is far from every change point in T^*. In other words, a smaller value of Hausdroff index is an indicator of an overall better performance.

3 Experimental Results and Discussions

3.1 Enviroment Setup

We have designed an experiment involving different forms of obstacles and moving objects to best collect the noise and other disturbances that IoT devices may face when attempting to profile their neighboring nodes using RSSI observations. In our experiment we have used two Digi XBee 3 Series programmable modules implementing IEEE 802.15.4. [14] (as depicted in Fig. 3), where one device acts as the temperature reading sensor transmitting its reading to the receiver. In this experiment, the transmitter and the receiver are situated in a residential property separated by a floor/ceiling and interior walls (depicted in Fig. 4). The 5 occupants living on the property have their routine daily schedule of moving around the property during the day (i.e., significant movement) and resting (i.e., minimal movement) at night.

Using an embedded python script in the receiving node, the RSSI values observed by the receiving node were reported to a workstation and subsequently stored. Refer to Fig. 1 for a visual depiction of this RSSI time series where significant change in the volatility of the timeseries can be observed between the two defined periods.

Then, the captured (i.e., stored) RSSI timeseries was processed using the *change point detection* schemes surveyed in Sect. 2. In our experiment, we have used python libraries (a) Raptures [8] and (b) Matrix Profile [15] to implement and assess effectiveness of the surveyed methods (from Sect. 2) on segmentation of the collected RSSI timeseries. The dataset and python codes of this work can be found on https://github.com/pooria121/rssisegmentation.

Fig. 3. Digi XBee 3 Series programmable module implementing IEEE 802.15.4. in a weatherproof secure enclosure protecting the devices from the elements when deployed.

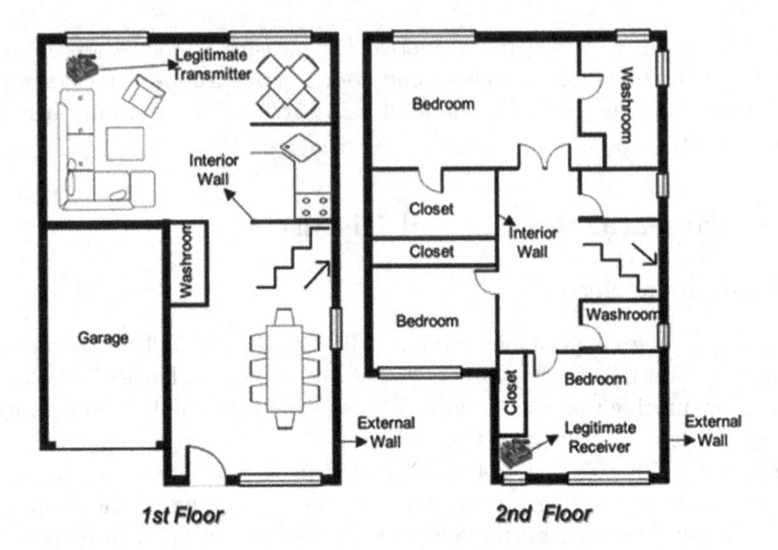

Fig. 4. The legitimate transmitter is situated in the first-floor family room while the legitimate receiver is situated in the second floor's bedroom separated by interior walls and an interior floor. The 5 occupants in the property are considered to be the influencing moving objects.

3.2 Results and Discussions

The summary of segmentation performance of surveyed schemes in Sect. 2 in respect to RandIndex and Hausdorff criterions are provided in Table 1. Also, visualizations for segmentations of each of the cost functions and search methods are provided in Appendix A. Note that some of the cost functions are not compatible with some of the search methods. As a result, certain cells in Table 1 are left blank to reflect such incompatibility.

From Table 1, we can observe effectiveness of ℓ_1 cost function for change point discovery in RSSI timeseries. This cost function performed well irrespective of search methods used. It is worth noting that all the surveyed search methods (except for Dynamic Programming and Matrix Profile) are approximating the solution to (3) and yet their segmentation performance is on par with the dynamic programming approach (which attempts to find the exact solution to Eq. 3). This observation is particularly important since runtime complexity of the approximating search methods is much lower than that of the dynamic programming. In other words, for long RSSI timeseries, the approximating search methods can be as effective or even faster in finding an acceptable segmentation.

Table 1. Summary of segmentation performance of surveyed methods applied to the collected RSSI time series.

	Linear		ℓ_1		ℓ_2		Gaussian		Autoregression		Cosine	
	RandIndex	Hausdorff	RandIndex	Hausdorff	RandIndex	Hausdorff	RandIndex	Hausdorff	RandIndex	Hausdorff	RandIndex	Hausdorff
Dynamic Programming			0.999	23	0.678	13888	0.998	102	0.511	29283		
Window Sliding			0.524	41908	0.524	41908	0.502	36153	0.515	40438		
Binary Segmentation	0.678	13888	0.999	23	0.678	13888	0.998	102	0.511	29283	0.528	42406
Bottom Up	0.528	26302	0.999	23	0.678	13888	0.981	672	0.516	40438		
Matrix Profile											0.999	67

Also, it is worth noting that the Matrix Profile is attaining "almost" the same segmentation performance while being fundamentally different than all the other experimented search methods. As surveyed in Sect. 2, the Matrix Profile performs segmentation as a side-effect of its data structure that it uses to identify and store similar data points in a timeseries. Recall that Matrix Profile segmentation complexity is a linear function of the number of the data points in the timeseries, which makes it a noteworthy item in Table 1. Such low overhead cost during the runtime would makes this approach a prime candidate for implementation on IoT nodes with limited computational resources (e.g., battery capacity and processing power).

4 Conclusions and Future Work

In our experiments we have successfully demonstrated that RSSI timeseries of a typical IoT node contains different regimes with different statistical (e.g., mean and variance) properties due to variabilities, such as the presence of moving obstacles in a typical IoT environment. Therefore, detecting change points in a given RSSI timeseries, in order to find regions with different statistical properties, can only enhance the robustness of the modelling process that may follow.

Time series segmentation is a widely studied field given that many real-world datasets, such stock market prices, heart electrocardiogram (ECG), and sensor readings are all timeseries data. As a result, many cost functions and search methods have been formulated to address a variety of applications requiring time-series segmentation at their core. In our research, we have successfully attempted to study some of the best known timeseries segmentation search methods and well-recognized cost functions in order to better understand their effectiveness against RSSI timeseries. We have demonstrated that the ℓ_1 cost function can capture a meaningful relationship between neighboring data points in a RSSI timeseries and result in a stable segmentation across different search methods.

Profiling wireless IoT nodes using their RSSI signature pattern as captured by a central station is used in many cybersecurity applications such as identity authentication and MAC spoofing detection [5]. Because of the possibility of adversarial contamination [16], any use of RSSI samples, including at the segmentation stage, must be carefully planned. As a natural extension of this work, it is necessary to study robustness of change point detection schemes under adversarial noise. In other words, a carefully crafted adversarial noise introduced by a malicious actor into a dataset of RSSI values could affect the segmentation performance. As our future work, we plan on furthering our understanding and experimentation to better understand the use of RSSI timeseries segmentations in an adversarial online setting.

Appendix A

Segmentation visualization of algorithms reported in Table 1 (Figs. 5, 6, 7, 8, 9 and 10).

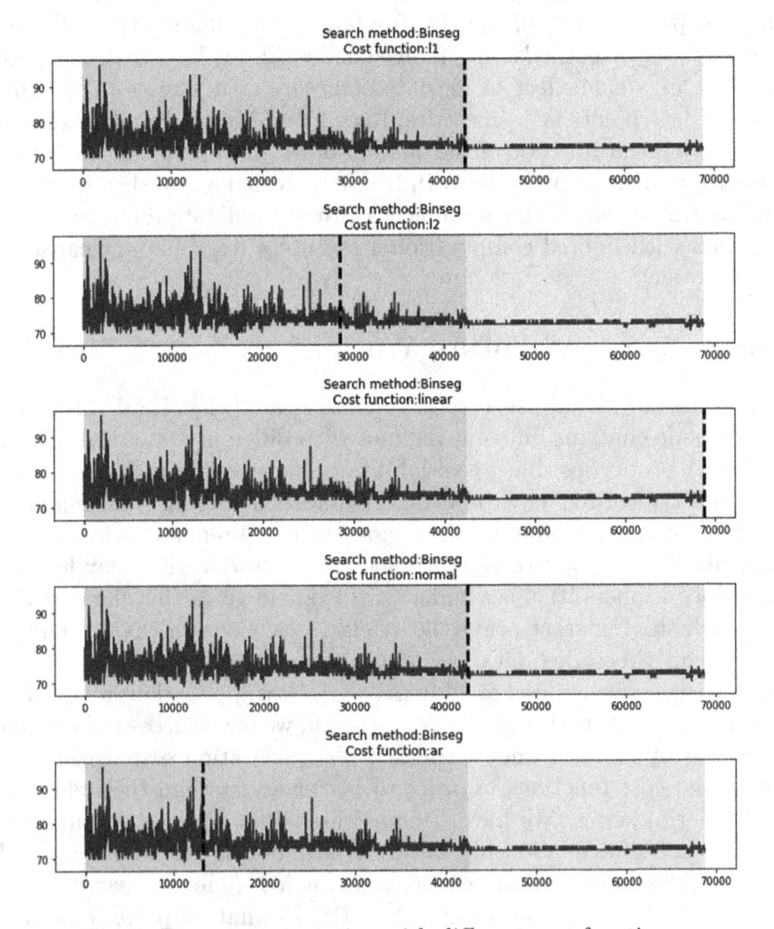

Fig. 5. Binary segmentation with different cost functions.

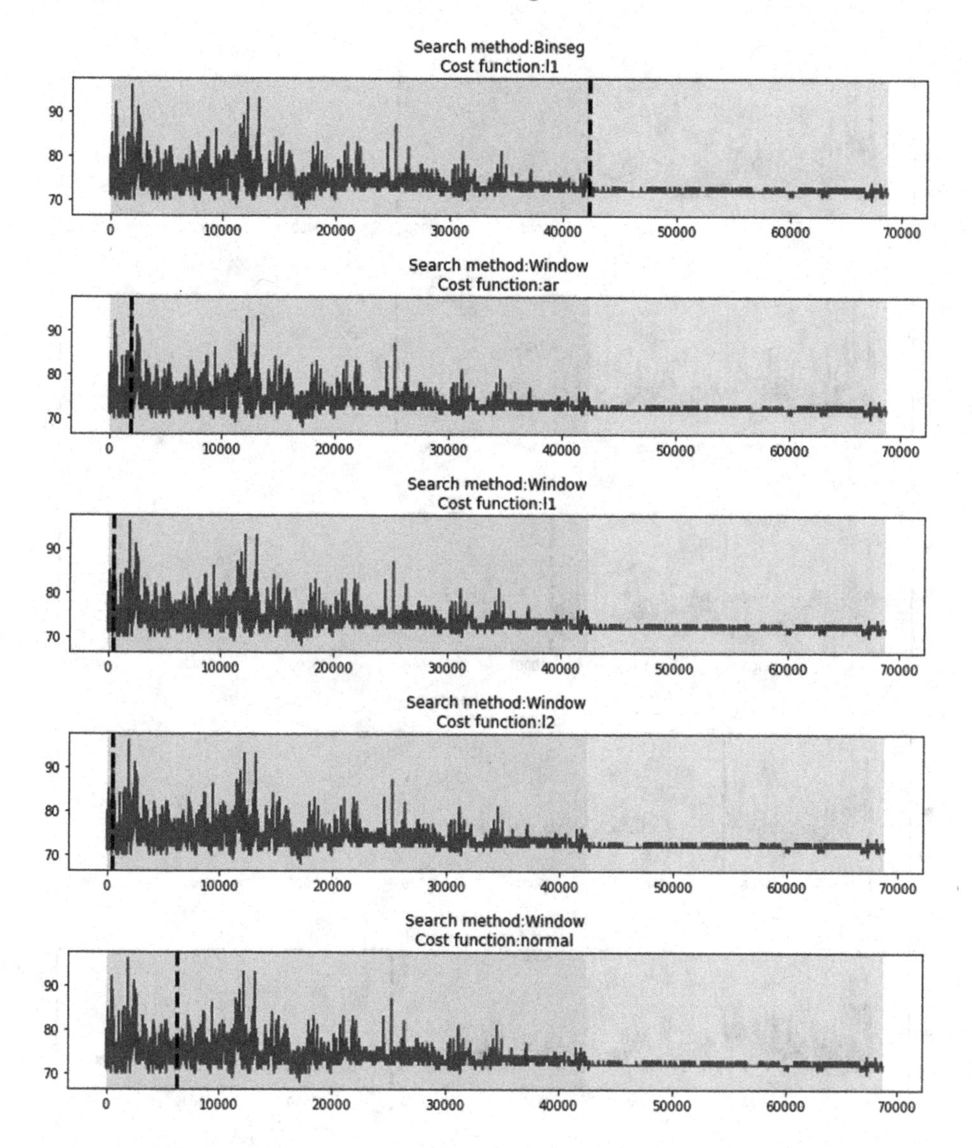

Fig. 6. Windowing segmentation with different cost functions.

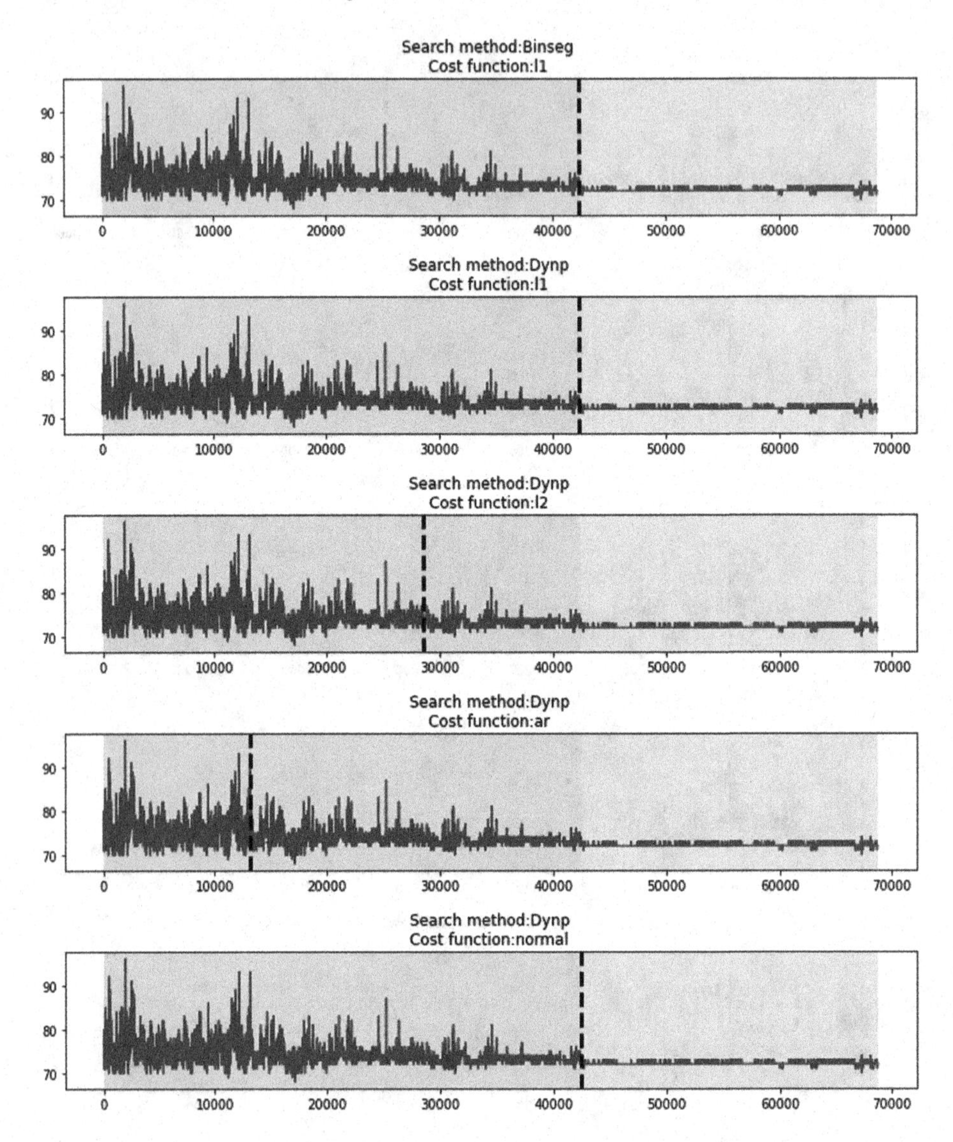

Fig. 7. Dynamic programming segmentation with different cost functions.

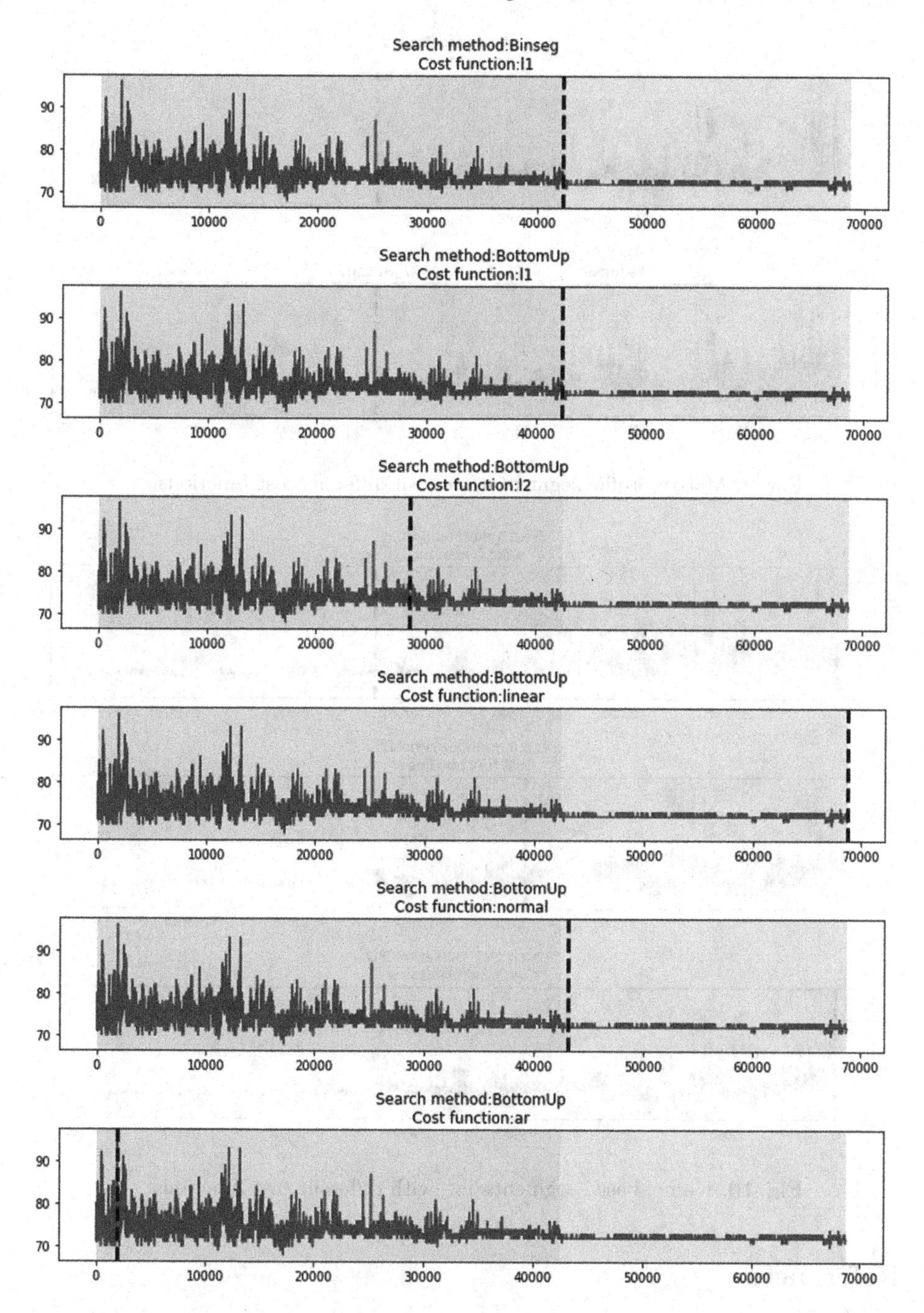

Fig. 8. Bottom-up segmentation with different cost functions.

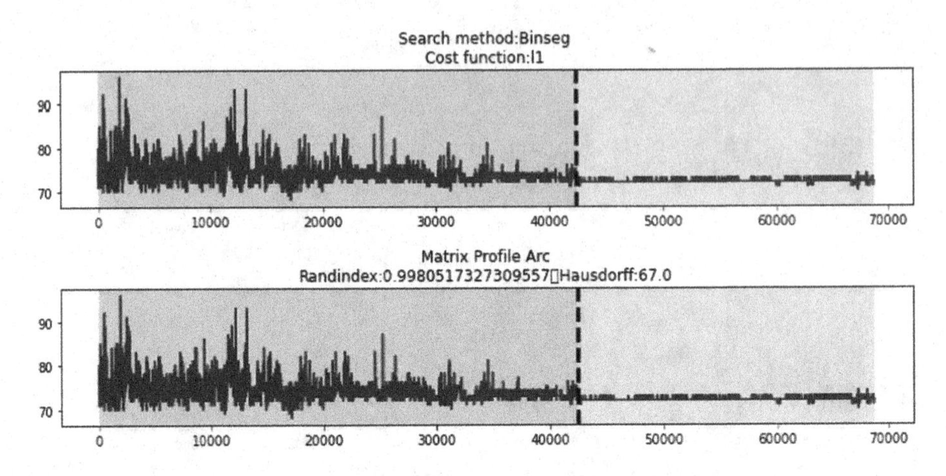

Fig. 9. Matrix profile segmentation with different cost functions.

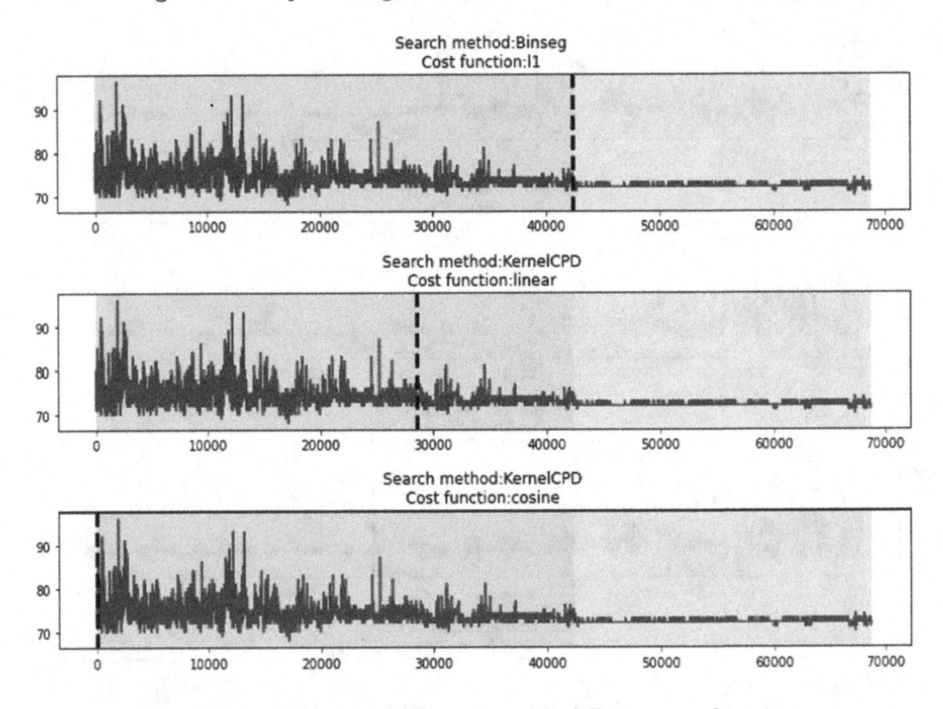

Fig. 10. Kernel-based segmentation with different cost functions.

References

1. Wenjia, W., Xiaolin, G., Dong, K., Shi, X., Yang, M.: PRAPD: a novel received signal strength-based approach for practical rogue access point detection. Int. J. Distrib. Sens. Netw. **14**(8) (2018). https://doi.org/10.1177/1550147718795838

2. Moosavirad, S.M., Kabiri, P., Mahini, H.: RSSAT: a wireless intrusion detection system based on received signal strength acceptance test. J. Adv. Comput. Res. **4**(1), 65–80 (2013)
3. Demirbas, M., Song, Y.: An RSSI-based scheme for sybil attack detection in wireless sensor networks. In: 2006 International Symposium on a World of Wireless, Mobile and Multimedia Networks (WoWMoM 2006), pp. 5-pp. IEEE (2006)
4. Madani, P., Vlajic, N., Maljevic, I.: Randomized moving target approach for mac-layer spoofing detection and prevention in IoT systems. Digital Threats Res. Pract. (2022)
5. Madani, P., Vlajic, N.: RSSI-based MAC-layer spoofing detection: deep learning approach. J. Cybersecur. Privacy **1**(3), 453–469 (2021)
6. Sandeepa, C., Moremada, C., Dissanayaka, N., Gamage, T., Liyanage, M.: Social interaction tracking and patient prediction system for potential COVID-19 patients. In: 2020 IEEE 3rd 5G World Forum (5GWF), pp. 13–18. IEEE (2020)
7. Sugano, M., Kawazoe, T., Ohta, Y., Murata, M.: Indoor localization system using RSSI measurement of wireless sensor network based on ZigBee standard. Wirel. Opt. Commun. **538**, 1–6 (2006)
8. Truong, C., Oudre, L., Vayatis, N.: Selective review of offline change point detection methods. Signal Process. **167**, 107299 (2020)
9. Frick, K., Munk, A., Sieling, H.: Multiscale change point inference. J. R. Stat. Soc. Ser. B (Stat. Methodol.) **76**(3), 495–580 (2014)
10. Nam, C.F.H., Aston, J.A.D., Johansen, A.M.: Quantifying the uncertainty in change points. J. Time Ser. Anal. **33**(5), 807–823 (2012)
11. Yeh, C.-C.M., et al.: Matrix profile I: all pairs similarity joins for time series: a unifying view that includes motifs, discords and shapelets. In: 2016 IEEE 16th International Conference on data mining (ICDM), pp. 1317–1322. IEEE (2016)
12. Lajugie, R., Bach, F., Arlot, S.: Large-margin metric learning for constrained partitioning problems. In: International Conference on Machine Learning, pp. 297–305. PMLR (2014)
13. Harchaoui, Z., Lévy-Leduc, C.: Multiple change-point estimation with a total variation penalty. J. Am. Stat. Assoc. **105**(492), 1480–1493 (2010)
14. Safaric, S., Malaric, K.: Zigbee wireless standard. In: Proceedings ELMAR 2006, pp. 259–262. IEEE (2006)
15. Van Benschoten, A., Ouyang, A., Bischoff, F., Marrs, T.: MPA: a novel cross-language API for time series analysis. J. Open Source Softw. **5**(49), 2179 (2020)
16. Madani, P., Vlajic, N.: Robustness of deep autoencoder in intrusion detection under adversarial contamination. In: Proceedings of the 5th Annual Symposium and Bootcamp on Hot Topics in the Science of Security, pp. 1–8 (2018)

Consumer-Friendly Methods for Privacy Protection Against Cleaning Robots

Yanxiu Wuwang(✉) and Gunther Schiefer

Karlsruhe Institute of Technology, Karlsruhe, Germany
yanxiu.wuwang@student.kit.edu, gunther.schiefer@kit.edu

Abstract. As Internet of Things (IoT) technologies enter the consumer market, smart cleaning robots have gained high attention and usage in households. However, as the "privacy paradox" phenomenon states, consumers behave differently even if many claim to be concerned about smart robot privacy issues. In this paper, we describe our attempt to discover effective measures for average consumers to guard against potential privacy intrusions by cleaning robots. We define our target devices, provide an ideal smart home network topology and establish our threat model. We document network redirection and analytic methods we used during our research. We categorize existing privacy protection methods and describe their general procedures. We assess and evaluate the protection methods with regard to three aspects: protection effectiveness, functionality loss and consumer-friendliness. In the end we perform a tabular qualitative comparison and develop our vision for privacy protection against cleaning robots.

Keywords: Cleaning robot · Privacy · Protection · Consumer · Internet of things

1 Introduction

A trend in the modern development of digitalization is the usage of IoT devices. IoT devices, according to [31], are defined as forming "the ubiquitous and global network that helps and provides the functionality of integrating the physical world." Although rooted as part of the "Industry 4.0" revolution, IoT began to gain great attention in the consumer market in recent years. Smart robots are now becoming a common household IoT device: According to Bitkom [10], 18% of all interviewees already use a robot in their homes. While more categories of home robots appear on the market every year, cleaning robots remain the dominant robot type used by households. 67% of the interviewees are interested in or already using a (ground) cleaning robot [10]. They are affordable, easy to use, and can be controlled anywhere from the world by using a cloud service together with a mobile device.

Smart home robots, including cleaning robots, could potentially create a severe threat to user privacy. Such threats could be informational or territorial,

© The Author(s), under exclusive license to Springer Nature Switzerland AG 2022
W. Li et al. (Eds.): ADIoT 2022, LNCS 13745, pp. 102–121, 2022.
https://doi.org/10.1007/978-3-031-21311-3_5

creating observable influence in the users' physical environment [26]. A cleaning robot connected to a home network could perform local network discovery (such as using ARP sweeps) and collect data about other devices in the same network. Cleaning robots equipped with modern technologies like Light Detection And Ranging (LiDAR) navigation could be further manipulated to eavesdrop people without being noticed [33].

Another empirical phenomenon that is highly related to consumer privacy behaviour is the "Privacy Paradox" [30], that is, "Although consumers seem to be concerned about their privacy ... their behaviours diverge from their intentions to disclose personal details." Statistics from the European Commission in 2010 [18] support this theory. Consumers understand the importance of privacy protection, but such protection measures, in general, can not be taken without an adequate level of understanding of the technologies behind online services. Therefore, the practical relevance of developing effective protection measures tailored for consumers cannot be overstated, especially for a relative new technology field like IoT and cleaning robots. We set out the following research question:

What are privacy protection methods against smart cleaning robots and are they suitable and effective for average consumers?

The paper consists of 6 sections. In Sect. 2, we define basic concepts, provide a network topology for typical smart home setups and a threat model for privacy intrusions. We briefly introduce our used methodologies in Sect. 3. In Sect. 4 we describe the general ideas and procedures for all privacy protection methods conducted in our research. Section 5 evaluates the effect and consumer-friendliness of the methods. We summarize our results and provide our view on the current state of available methods in Sect. 6.

2 Definitions and Research Models

This section defines necessary concepts, assumptions and the threat model for the research.

2.1 Definitions

Smart Cleaning Robot. A robot is a machine that acts as a human or takes over a task from a human [22]. Our research focuses on a specific type of robots: service robots that deal with household ground cleaning.

Specifically, our research is focused on smart cleaning robots. We use the definition from Könings et al. [25] and Weiser [39] in the ubiquitous computing context: the ability to support users in their activities by processing and interpreting context information in order to adapt to the user's needs. We describe the "smartness" of a cleaning robot considering two aspects:

- Smart cleaning robots process context information through its sensors.
  In order to determine its own location and clean a certain region, smart robots must be equipped with sensors to perceive information about its environment, for example, its relative position inside the room and distance to solid barriers. To achieve this, smart cleaning robots use collision sensors to detect walls and infrared sensors to sense solid objects from a distance. More advanced robots (like the one we used during the research) include a LiDAR sensor that may provide higher measurement precision [9]. Some high-end robots have a camera and could perform object identification through machine learning.
- Smart cleaning robots interpret context information through cloud servers with on-board computing assistance.
  While smart cleaning robots carry an on-board processor, most of them rely on a cloud server for some specific tasks, for example, supplementary tasks that assist the cleaning operation. Task distribution between the server and the robot itself is model-specific, but features like remote control could only be implemented with the assistance of a cloud server due to home networks normally being isolated under Network Address Translation (NAT).

Smart Mobile Device. A smart cleaning robot can be configured and remotely controlled by a smart mobile device. A smart mobile device is a portable electronic device that enables its users to access a variety of local and remote services [35]. This includes smart phones and smart tablets, mostly running iOS or Android [34].

An accompanying app can be installed on the smart mobile device to control the robot. The app sends network requests to a vendor-controlled cloud server and the server then notifies the robot to perform corresponding actions. Typical actions include control commands (start cleaning, pause cleaning, return for charging) and device configurations (map configuration, network configuration).

The rest of this paper does not distinguish different types of smart mobile devices. Terminologies including "smart device" "smartphone" "mobile device" are treated as synonyms.

2.2 Research Devices

Smart Cleaning Robot. The device model we used throughout the entire research process is Dreame Robot Vacuum D9 Pro[1], a cleaning robot produced by the Chinese company Xiaomi, released under the Dreame brand. The community-maintained database `dontvaccum.me` [14] provides detailed technical specifications on this device: it has a LiDAR sensor, infrared sensors and water pumps. This device has no camera and thus does not have object identification functionalities. As a result, user privacy threats originating from optical object identification algorithms cannot be assessed in our research.

[1] This exact device model is no longer officially listed. The current variant "Dreame D9 Mistral Pro" [16] only differs in colour.

Smart Device. All Xiaomi smart devices, including our target robot, use the "Xiaomi Home" app [42] for device management. The app requires a Xiaomi account to login; after logging in, the user can manage all Xiaomi smart home products associated with this account. For our cleaning robot, the app offers two one-key control buttons (start cleaning, return for charging) and a map management user interface.

In our research we used two iOS devices: one iPhone 7 running iOS 14, jailbroken (root privilege acquired for the operating user) and one iPhone SE (2nd generation) running unmodified iOS 15. Both of them provide the same functionalities while using the app. Some operating system-specific features are discussed below since they assist the users in protecting their privacy. Note that the root privilege is only used in the research process to gain more insight into the application logic and data transmission; none of our proposed protection methods requires such technical modification. We did not look at Android devices and therefore could not research on Android-specific features, but the app itself should function similarly and communicates with network services in the same way as on the iOS platform.

2.3 Network Topology

In a typical home setup (like the one described in the robot setup manual [15]), the smart device and the cleaning robot are connected wirelessly (IEEE 802.11) to a home network. The following network topology (Fig. 1) is used in our research and resembles a minimal smart home environment:

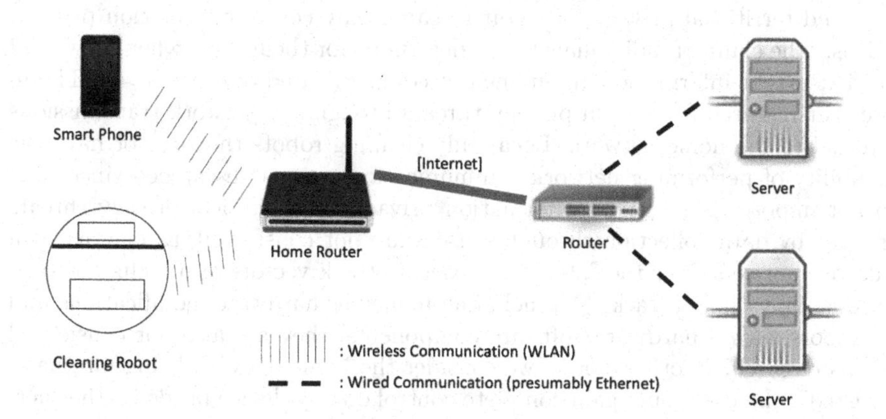

Fig. 1. Network topology of a typical home network

The left hand side of the figure is the home network. Home networks typically use an "all-in-one" home "router" (for instance, the Vodafone Station, [38]) that acts as a switch, a router, a wireless Access Point (AP) and a firewall simultaneously. The home router also bundles server functionalities like DHCP server and DNS server.

The robot and the smartphone are connected to the home AP wirelessly and assigned private IP addresses under the same subnet, in our case 192.168.0.0/24. Mesh deployment is not considered and the home router acts as the default gateway.

The home router transmits data through the Internet. Any intermediate routers could theoretically sniff and analyse data, which is an important part of our threat model in Sect. 2.4.

The right hand side of the figure represents cloud servers. We assume that any number of cloud servers could be used to handle data coming from the home network. Such cloud servers process requests and send responses over the Internet back to the home network.

Note that some of the protection methods introduced in Sect. 4 change the network topology given in this section, for example, by introducing an additional router between the robot and the home router.

2.4 Threat Model

In this section we describe aspects of privacy threats we care about in this research and the possible attack surfaces we assume.

Two types of privacy are distinguished in existing literature: information privacy and territorial privacy [26]. Our research only covers information privacy, that is, "the claim of individuals [...] to determine for themselves when, how, and to what extent information about them is communicated to others" [40]. Therefore, our research focuses on privacy threats brought by network transmissions external to the home network. Local-only cleaning robots that do not have the capability of performing network communication are not researched since they do not impose risks against information privacy. Sole physical privacy threats brought by data collection through sensors are not considered, as long as such data are not sent over the Internet. Physical attack vectors also include a variety of supply chain attacks [27] including malicious hardware modifications and backdoors within hardware/software components; they are also not considered in this research. In other words, we consider the home network being internally "trusted": the user could gain complete control on any device inside his/her network. As a consequence, data that never leave the home network are not treated as a privacy risk.

We only start to consider threats of information privacy when traffic leaves the default gateway. From the first routing hop to the final transmission destination, we assume that every intermediate router could potentially sniff and analyze data. We also do not restrict the capabilities of Man-in-the-Middle (MITM) attacks: all intermediate nodes could theoretically read packet headers and data in any layer and also perform Deep Packet Inspection (DPI) to gain further insights into the payload data. Since DPI utilities are widely available, highly effective and built-in into many vendor devices [12], such assumption is justified.

When data arrive at the vendor's site, cloud servers would have to decrypt the data. Thus, encryption should not be considered as a protection mechanism against the vendor. Privacy threats imposed by the cloud are restricted to the data it received; we exclude any other possibilities (like time-based side channel attacks) that could allow the vendor to gain user information outside network transmissions.

We also assume that the cloud servers behave "passively": they only react on user command during the operation of the robot and the smart device. The vendor could neither actively control the robot (without users' consent) nor the smartphone, turning both devices into a trojan. The robot is also assumed not to carry any additional actuators except the ones required for cleaning operations like a rotor for mopping or water pumps, excluding any physical possibilities of vendors to willingly spy on the users.

Different types of data have different levels of sensitivity and bring unequal threats to users' privacy. Data types discussed in this research are wider than General Data Protection Regulation (GDPR) [17] "personal data" definition: some fingerprint data, including device identifiers and vendor codes, are pseudonyms and do not always constitute personal identification, but they could be used in an aggregation process to recover personal attributes and increase identifiability. Anonymous data like hardware types and versions generate minimal risks and are, in some cases, necessary for normal operation, but they could still be used, in aggregation with other identifiers, for marketing and personalization purposes and thus remain part of our threat model. Therefore, reducing the amount of data transmitted would be the most direct purpose and effect of many of our methods discussed below.

3 Methodologies

We applied various assumed privacy protection methods on the robot and compared data transmitted to cloud servers with and without such protection methods. This section introduces tools for relevant data capture and analysis conducted.

3.1 Traffic Redirection

In order to analyse any network traffic, the observation device (our laptop) must be an intermediate node of the transmission, acting as a man-in-the-middle.

In this research, we used the application proxy Mitmproxy [28] on the smart device. We could not configure a proxy server on the robot since there is no configuration interface available. In this case, we used a Raspberry Pi 3 as a custom router. Adding the custom router changes the network topology during the research process. Figure 2 represents the network topology of the home network with our custom router added during the research process. During robot initialization, we instruct the robot to bind itself to the new custom router instead of the outer home router. Software for traffic analysis is also installed on the custom router. This allows us to observe traffic from the robot using the observation device even without access to the system on the robot itself. Note that the custom router is only used for data collection during the research and does not resemble part of a typical home network.

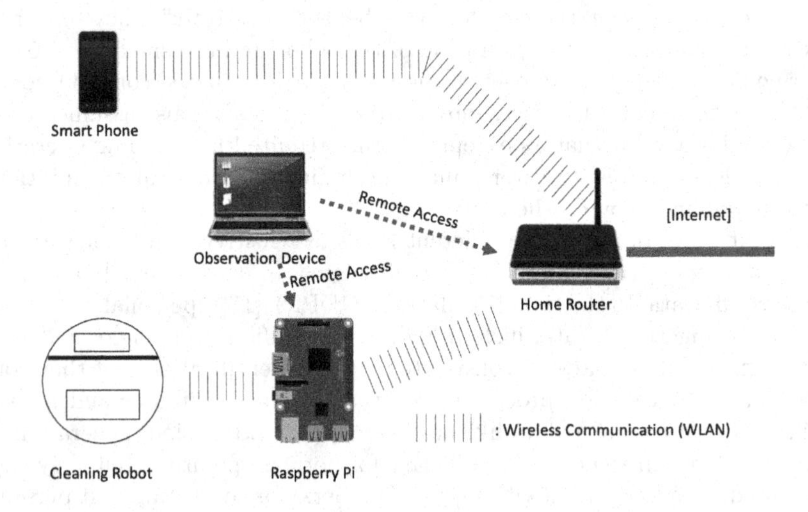

Fig. 2. Network topology with custom router

3.2 Traffic Analysis

After observing traffic from the smart robot and the smart device, data must be collected and saved for further analysis.

Wireshark [41] is used as the main network protocol analyser. For known HTTP(S) data, Mitmproxy has a built-in protocol analyser and provides better insight than Wireshark. The custom scripting feature of Mitmproxy also allows us to write programs that enable further assisted processing, such as data decryption of a non-standard protocol suite.

Note that in order to decrypt Transport Layer Security (TLS)-encrypted traffic, a custom certificate must be installed and trusted by the smart device. Mitmproxy then performs a certificate sniffing, replacing the upstream certificate by the ones generated by Mitmproxy itself. This is not possible for the cleaning robot; we could not access the operating system of the robot. Therefore, we could not decrypt and obtain the data payload sent from the robot to remote Application Programming Interfaces (APIs). Our observation is restricted to data extracted from packet headers, including domain names and protocol negotiations.

3.3 App Analysis

On a full-fledged desktop operating system, application data could be easily extracted from the filesystem. This is not the case for mobile operating systems, especially on iOS where the system is heavily locked down and normal users cannot access the filesystem directly.

To extract data from the Xiaomi Home app, we used publicly available tools [24] to gain superuser privileges (known as "jailbreaking") on the smartphone and then extracted data from the sandbox directory of the Xiaomi Home app. In this research we did not perform any binary reverse engineering on the app itself; we only explored saved data inside filesystem directories.

4 Privacy Protection Methods: Procedural Description

In this section we enumerate all privacy protection methods we researched. We categorize all methods into 3 types (isolation, restriction, specialized hardware) and describe their general procedures in each subsection.

4.1 Isolation Methods

Isolation methods segregate the smart robot and/or the smart device from other devices on the local network, removing possibilities that home devices communicate with each other or extract information about other devices. Note that without additional restriction methods, isolated devices could still communicate with the Internet.

Robot and Smart Device: Network Isolation. Through a simple ARP sweeping, the robot could gain abundant fingerprint information including device types and hardware identifiers on all devices in the same network. The idea of network (subnet) isolation is to create a separate subnet for the robot. No other devices connect to this separate subnet. As a result, the robot could lose the potential to detect other devices in the same network.

This protection method is deployable on home routers due to the "guest network" functionality. This feature is originally intended to isolate untrusted guest devices from regular home devices. Here we could use it to create a new robot-specific network. Note that different routers may provide different degrees of isolation despite all claiming that they create a "new" network on its interface; it may or may not actually create a new subnet.

Network isolation can also be configured for the smart device.

Smart Device: App Isolation. Another isolation mechanism is to install the application on a separate smart device. In other words, the user does not use his/her ordinary device to control the robot; instead, a specialized device, sometimes ambiguously called a "smart controller hub" by IoT vendors, is used.

Such devices are not intended to be used for any other purposes. Only apps related to smart home management (in our case the single Xiaomi Home app) are installed on the device.

The main purpose of such method is to prevent the app from reading other sensitive information on the device, especially data from other apps. Both Android and iOS have sandboxing mechanisms (Apple Platform Security, [6]; Android Application Sandbox, [2]) built-in, but as we investigate further in the next section, such protection could have positive effects on privacy.

4.2 Restriction Methods

Methods introduced in this section attempt to forbid certain Internet communications.

Robot: Disconnection. The most extreme form of network restriction is to completely disconnect the robot from the Internet. According to our threat model, this always provides maximum protection, since no data could be sent over the Internet.

Numerous hardware and software methods exist to cut Internet communication beyond physical disconnection. Some other protection methods, despite not disrupting all Internet connections or not intended to do so, eventually create the same effect as a whole network disconnection. The exact side-effects and difficulty of performing such disconnection would depend on the concrete procedure used.

Since mobile smart devices are supposed to be always-on and multi-purpose (with the exception of app isolation mentioned above), a complete Internet disconnection of the smart device is not considered as a viable solution.

Robot: Domain Restriction. A more gradual approach is to selectively forbid Internet connections based on a set of certain rules. In our research we explored domain-based Internet restrictions. More granular restriction methods, including URL-based pattern matching, require a DPI implementation and are not deployable for consumers. We used the popular advertisement blocker PiHole [32], since it is domain-based and easily configurable. Some router vendors provide extensive documentation (like ASUS, [7]) on how to use PiHole with its routers, while others (for instance, [11]) come with modifiable, extensible firmwares out-of-the-box which make a PiHole setup easier.

Smart Device: Domain Restriction. PiHole could also be configured for the smart device.

Internet communications on the smart device happen much more frequently than on the robot. The cleaning robot is a single-purpose device that only communicates with several dedicated APIs; on the other hand, multiple apps and services could run simultaneously on the smart device, each communicating with its own set of APIs, adding noise packets during our research. In our research with iOS devices, we filtered out Apple platform-related API calls and focused domain restrictions on either Xiaomi-related domains or external tracking services.

Smart Device: Permission Configuration. Most smart devices nowadays provide granular permission configurations down to every specific application. Such configurations typically offer access restrictions to sensors and services including location services, cameras and microphones. iOS also allows restrictions on networking and Bluetooth. The underlying platform guarantees that the app cannot gather information from such sources if the corresponding permission is not granted. Since permission configuration affects functionalities of the application, an examination of the lowest possible permissions for an app is required.

Some smart devices, including iOS since version 15, also provide a report on the usage of all permissions for every installed app, known as the "App Privacy Report". The system keeps a log of all data, sensor & network access (endpoint statistics only on iOS) of all apps for the past 7 days and provides an interface to review all permission usage. iOS also allows such logs to be exported; we used an external analyser "App Privacy Insights" [23] to extract further details in our research.

4.3 Home Security Device

In the end we researched network devices that are specially branded as "Home Security Device" on the market. These devices typically introduce functionalities like network access restrictions, traffic monitoring and scheduling that may assist the user in protecting their information privacy.

Fingbox. We studied the product "Fingbox", version 1 [20] in our research.[2]

To integrate Fingbox into the home network, a wired Ethernet connection is required. A Fing account is also required to activate the device and to connect to Fing cloud services, allowing the user to monitor his home network anywhere. This behaviour might impose additional privacy threats to the user, but relevant impacts are not topic of this research.

Fingbox must be paired with the app "Fing" on the smart device. The device periodically scans all other devices within the same subnet, after which the user is allowed to perform management operations. We will discuss the effects of two specific operations, "Unblock Device" and "Pause Internet" [21] in the next section.

5 Privacy Protection Methods: Evaluation

This section evaluates all protection methods mentioned in the previous section.

[2] No substantial functional differences exist in the currently sold version 2.

5.1 Evaluation Framework

For every privacy protection method, we define our evaluation framework from three aspects:

Protection Effectiveness. A protection method must create a practical effect defending against unnecessary and sensitive data transmission. In our threat model, methods that allow less data transmission provide better protection than methods that allow more; methods that block sensitive information are more effective than those only blocking anonymous information. Also, due to possible MITM sniffers, methods that block or filter unencrypted traffic provide better protection.

Functionality Loss. A protection method typically reduces or limits Internet communications, which may cause loss in functionality. We tried to search for protection methods that do not require a sacrifice on functionality. This is not always implementable as we will see in the following evaluations.

Functionalities can be categorized into various levels of importance. We consider a functionality as "important" or as a "core functionality" when the purpose of the product (the robot) could not be realized without such functionality. The main purpose of a cleaning robot is vacuuming the ground, which we treat as the core functionality. Controls on starting and stopping the robot are also of high importance. Other functionalities that improve user experience on realizing the main purpose are considered supplementary, including remote management and map configuration. Methods that only reduce supplementary functionalities are better than those which affect core functionalities.

Consumer Friendliness. Our research focuses on protection methods that are implementable for average consumers. Specifically, we assume that the abilities of average consumers are as follows:

- Average consumers do not have a technical background. They cannot perform modifications that require technological understanding like DPI and device rooting.
- Average consumers cannot perform hardware modifications. For example, certain protection methods like installing a third-party firmware (Valetudo [37] being a common example) provides high protection effectiveness but requires flashing through UART serial port [13]. They are intentionally not discussed in the last section.
- Average consumers cannot perform complex software modifications. Methods that only require controlling devices through the given graphical/web user interface are considered consumer-friendly. Average consumers are assumed not to operate a command-line interface directly, but they might be able to use prepared tools with detailed documentation. Such methods are evaluated with low friendliness. Another factor of consumer-friendliness is the monetary cost

of deploying such methods. Methods bundled with lower costs (for instance, methods that do not require purchasing extra hardware) are more consumer-friendly than those that involve extra spending.

5.2 Isolation Methods

Isolation methods are effective if devices gather information through sensors and network activities, and send them to the cloud server.

Robot: Network Isolation. We did not observe any local network sweeping activities performed by this robot with the current installed firmware. The robot sends ICMP pings to the router periodically, which is necessary to perform online detection. Such behaviour might be added in a future update, and under this assumption network isolation could provide a certain level of protection.

In general, the method does not provide a high level of effectiveness (until now). Network isolation does not bring any functionality loss and guest network setups are consumer-friendly.

Smart Device: Network Isolation. We were able to separate network traffic generated by various apps on the smart device by looking at the User Agent field `Xiaomi Home/6.13.201.1 CFNetwork/1197 Darwin/20.0.0`. Like in the robot case above, we did not observe any local network sweeping activities with this app version. However, the app contacts a Xiaomi Router-specific API periodically, which on our non-Xiaomi router returns a `401` error.

iOS apps must go through the Apple App Store Review process before they become available and the Review Guidelines [4] impose strict requirements on information gathering and transmission. In Xiaomi Home app, this review process could potentially be circumvented, because the app uses hot updates. We discovered that the app downloads required code (for different home devices) on demand at runtime and stores them ("Plugins" stated by the app itself) under `Documents/Plugin` of the sandboxed data directory. A manifest is available at `Documents/PluginInfo/PluginInfoDict.plist`. Such plugins are downloaded directly from Xiaomi cloud servers and are not reviewed by Apple's App Review process. This behaviour greatly reduces restrictions on possible app behaviours and imposes huge forward risks to user privacy. Apple is also unlikely to forbid hot updates for all apps since such frameworks (like JSPatch, [8]) are heavily used by games to dynamically update logic and resources.

In general, network isolation does not provide huge effectiveness now but could be potentially threat-resistant in the future. Network isolation does not bring any functionality loss and guest network setups are consumer-friendly as the same case for robots.

Smart Device: App Isolation. iOS devices do not allow apps to read the list of installed apps. Ideally, such a restriction on the iOS platform has made this method unnecessary and no further protection effect could be achieved.

However, we discovered that the Xiaomi Home app abused an URL feature within iOS and performed a workaround. iOS supports one app opening another app through URLs by calling a function `[UIApplication openURL:options:completionHandler]`. Such schemes are registered under a description key `LSApplicationQueriesSchemes` [5] by each app individually. Therefore, one app could detect the existence of another app simply by checking if the corresponding URL exists through `LSApplicationQueriesSchemes`. Xiaomi Home registers 70 such detections, mostly being common Chinese apps including Taobao, Wechat, Kingsoft Office, etc. A full list could be viewed at our GitHub repository [43]. Since this is a built-in feature of the operating system and not a "vulnerability" on its own, the user cannot prevent such detection unless he/she applies the app isolation method.

In contrast, Android allows an app to read the list of installed apps as long as the app declares the corresponding permission [3] which simultaneously makes such behaviour visible from the user. App isolation could therefore bring further benefit to the user since no workaround is needed to read the list of installed apps on the Android platform.

The list of installed apps on a smart device is sensitive data and could be abused for behaviour analysis and targeted marketing. In general, isolating the Xiaomi Home app by installing it on a separate device avoids the effect of such detections and thus brings sufficient effectiveness. App isolation does not bring any functionality loss and is consumer-friendly except for the additional cost incurred of purchasing an additional device if necessary.

5.3 Restriction Methods

Robot: Disconnection. In our model, a full network disconnection always brings maximum effectiveness in protecting information privacy. Common methods include physical disconnection and router configurations. The former imposes no technical barrier, while the latter requires a certain degree of router control familiarity, for example, changing the AP password at minimum.

A full disconnection would forbid any remote management features; however, we discovered that the core functionality of our cleaning robot is not affected. Starting a cleaning process under an offline environment does not produce negative effects on cleaning, including advanced functionalities like LiDAR navigation and barrier avoidance. This also indicates that any network-based method cannot remove core functionalities from the robot, marking full disconnection the most extreme form of functionality loss in our research. This is not the case during the initialization process: the robot must be connected to the Internet to perform initial setup.

In general, a full disconnection brings both maximum effectiveness and maximum functionality loss. The method also achieves high consumer-friendliness.

Robot: Domain Restriction. Functionality loss and effectiveness depend on the chosen blocked domains. Since we could not decrypt API payload data, we could only deduct the purpose of a remote endpoint through its domain name and the observed effects created after blocking it.

Tracking domains are not responsible for any robot functionalities and can therefore always be blocked. Effectiveness would depend on the data transmitted through the tracking service. Due to not being able to observe transmission in clear text, we would assume that the robot might transmit any data, including all sensor information and fingerprint data. Our robot uses exactly one tracking service: Xiaomi's custom tracking system ("OneTrack") through domain `tracking.eu.miui.com`. Blocking this domain would bring significant effectiveness without any functionality loss.

Blocking other domains would make certain functionalities unusable. The robot uses 2 other APIs on the cloud server: one for map configuration and various system settings (domain `awsde0.fds.api.xiaomi.com`) and the other for remote control (domain `de.api.io.mi.com`). Blocking both creates the same effect as a full Internet disconnection. We discovered that blocking only the map management API does not disrupt remote control: on the smart device interface, the map disappeared but all control buttons remain usable. In other words, the 2 APIs work independently from each other.

One possible scenario that could bring a balance between functionality loss and effectiveness is that the user blocks the map API after finishing the entire configuration setup in the initialization phase. Since robot navigation does not rely on cloud servers, this method would prevent the cloud servers from receiving any data about the user's physical space, which could be highly sensitive, without losing the ability to control the robot through the smartphone.

In general domain restrictions provide adequate effectiveness with acceptable functionality loss. However, the difficulty of deploying PiHole is dependent on the home router. Despite having one-click deployment tutorials in the PiHole documentation [32], in many setups the user would be required to type in system commands on an unfamiliar interface. This method remains the most complicated method for average consumers among all of our methods researched.

Smart Device: Domain Restriction. The Xiaomi Home app uses multiple tracking services, including but not limited to Google Firebase, Facebook and Xiaomi's own tracking system ("OneTrack"). Through filesystem exploration, we could discover data logged by the tracking frameworks. We discovered that it performs a region difference and in extreme cases, OneTrack could log every single touch screen clicking, keystroke and changes to user interface navigation. These events are stored in a local SQLite database and periodically sent to the server. Local logging could not be turned off without extensive technical measures.

Domain blocking is possible for all such tracking services without losing any functionalities. In fact, since such tracking services are used by many apps, PiHole blocks corresponding domains by default without downloading external blocklists. For Xiaomi devices, additional blocklists that filter Xiaomi-specific domains are also available online [36].

In comparison to the cleaning robot, the smart device uses one single API (domain `de.api.io.mi.com`) for all communications (except Xiaomi account management and third-party logins). Blocking this domain produces the same effect as a full Internet disconnection. The API uses standard HTTP(S) but encrypts all data with a custom RC4 encryption-based protocol. We wrote a script ourselves [44] to parse traffic from Mitmproxy to HAR format and decrypt the API calls based on existing code [1] with modifications. We could then observe that the smart device sends not only fingerprint data but also location information, potentially including geographical data. Manipulating such data transmission is impossible without advanced MITM utilities.

Most traffic sent by the mobile device are indeed RC4 encrypted (which rules out the risk of intermediate router sniffing), but we noticed that in the initialization phase the app uses the XMPP protocol [29] to perform a handshake. The handshake transmits device identifiers and fingerprint data in plaintext without encryption. We did not discover the exact purpose of this handshake; but we observed that such a handshake is no longer necessary after initialization, so this XMPP domain could then be blocked to ensure all traffic to be encrypted properly.

In general, domain restrictions on mobile devices create high effectiveness against tracking services, but complete prevention of unwanted information transmission could not be achieved without maximum functionality loss. No API separation could be made and therefore the user faces an all-or-nothing choice. Also the method is not consumer-friendly due to the difficulty of the PiHole setup, as in the robot case above.

Smart Device: Permission Configuration. The Xiaomi Home app is aggressive on requesting various permissions. Furthermore, the app uses given permissions more often than the user would have expected, especially the location permission: App Privacy Report reports that the app requests location info twice every minute on average without visible user interface hints when running.

However, we discovered that turning off all given permissions (except networking) does not affect any functionalities of the application and the cleaning robot. This is also theoretically deductible: the app does not communicate with the robot directly (only through cloud servers) so local network permission is not required; the robot does not have Bluetooth communication capability; the robot also does not rely on location coordinates for navigation. Microphone permission is redundant since the system built-in voice assistant, Siri, could also be used for remote controlling.

In general turning off all permissions provides great effectiveness, especially against coordinate information without functionality loss. Such system configurations are directly built inside the operating system and are user-friendly.

5.4 Home Security Device

Fingbox. The effects of both Fingbox functionalities, "Block Device" and "Pause Internet", are exactly the same as a full Internet disconnection: the cleaning robot could not make any communication with the cloud servers; also no advanced filters could be configured on Fingbox. Therefore they achieve the same level of effectiveness and functionality loss as a full disconnection.

In general, the Fingbox method is more consumer-friendly to use than other disconnection mechanisms. The mobile app provides a one-click user interface to perform disconnection instead of performing a possibly complex router configuration. Moreover, Fingbox provides a "scheduled downtime" feature so that the user could configure to only disconnect the robot at specific time intervals. The downside is that the user would have to purchase such a device.

6 Conclusion

6.1 Comparison Table

Table 1 compares all protection methods described in Sect. 4 against the three criteria discussed in Sect. 5. Each criterion is evaluated with "Low", "Medium" or "High" for every method. A criterion evaluated with an interval (such as "Low-Medium") means that it is dependent on certain external factors, most typically including domains chosen to be blocked and the concrete Internet disconnection methods. Note that the "Low" "Medium" "High" values are not quantified; they are only qualitative assertions based on our discoveries. Developing a quantitative framework for asserting information privacy risks is also not part of our research purpose.

Table 1. Protection methods comparison

Name	Effectiveness	Functionality loss	Consumer friendliness
Robot network isolation	Low	None	High
Robot disconnection	High	High	Low-High
Robot domain restriction	Medium-High	Low-Medium	Low
Smart device network isolation	Low	None	High
Smart device app isolation	Medium	None	Low (with monetary cost)
Smart device domain restriction	Medium-High	Low-Medium	Low
Smart device permission configuration	Medium	None	Medium
Fingbox	High	High	High (with monetary cost)

We noticed two conflicting criteria among all methods above: the balance between effectiveness and functionality loss, along with the balance between effectiveness and consumer-friendliness. A method that provides better privacy (improved protection) results in either less functionality or higher deployment difficulty (low friendliness) or even both; however, a method that is easy to deploy and retains most functionalities typically produces little effect. Constrained to methods we discussed in our research, no perfect method exists that simultaneously provides high protection, little functionality loss and high friendliness.

6.2 Limitations

We acknowledge that our research has its own limitations.

Our research models are ideal and simplified. We assumed a simple home network with only one router, one smart device and one cleaning robot. We did not consider other IoT devices; interactions between smart devices (for example, triggering the cleaning robot from a smart door lock) could reveal more information about the user and create further privacy risks. Our threat model is simplified and considers only information privacy. Territorial risks imposed by physical intrusions (for example, by using an actuator on the robot) are not discussed.

Our methodologies are constrained and could not reflect the complete state of data collection by the target devices. This primarily happens on the robot side due to encryption and lack of attack surface of the system on the robot; we have managed to decrypt data sent from the smart device.

Our research devices are constrained and might not reflect the situation for all smart devices and smart cleaning robots. We only researched one specific robot model and its behaviour might not match robots with a different hardware model or produced by a different vendor. Only iOS devices are observed throughout the research and such behaviour might not be representative for an alternative mobile platform like Android. Protection methods beyond the ones proposed by us could exist when a different combination of home router, smart device and cleaning robot is used.

However, we believe that the result of our research remains relevant and could contribute to existing researches. Definitions given in Sect. 2 are general, containing a variety of devices beyond the target devices we used during our research. Theoretically, our research methodologies could be applied to any devices that match our definitions given above, which could serve as a starting point for studying potential protection methods for other devices. Functionalities of different cleaning robots and smart devices are similar; therefore, protection methods proposed by us could also be adapted for other devices, possibly with minor modifications (for example, changing the domain name to be blocked). Methods that do not depend on a certain vendor or platform, including disconnection and hardware home security device, could be reused in any home environment regardless of hardware type.

6.3 Current State of Privacy Protection

We conclude that the current state of available privacy protection methods for average consumers against cleaning robots is not mature and requires more research and development. There are two main aspects: the lack of vendor support and consumer technologies.

First, the vendors are not providing comprehensive control for the users to configure the cleaning robot regarding privacy. Article 24 of the GDPR [17] mentions two important concepts: privacy by design and privacy by default. From a technical perspective, the cleaning robot and the accompanying mobile app are mostly black boxes towards an end user. They are designed to be highly integrated so that the user has no control over individual functionalities like turning off specific sensors, not using specific cloud services, etc. The app asks for many permissions despite hardly any of them being necessary. From our perspective, this is far from a privacy-oriented product with a default privacy-friendly configuration.

Second, consumer-friendly privacy protection technology is still in its infancy. The European Data Protection Supervisor promotes the development of Personal Information Management Systems (PIMS) [19], but under the context of smart robots, such management systems are less likely to perform an efficient data transmission control due to the underlying encryption. Note that we are not promoting any concept of removing transmission encryption; they remain the most fundamental tools against privacy risks. But a PIMS in a cleaning robot context should allow users to inspect their own data and provide options to choose which data to transmit or not. Such systems, even for technical experts, could only be built with extensive MITM filtering tools which are, under the current state of technology, completely undesirable for average consumers. A "one-stop-shop" privacy management system remains highly theoretical until today.

Together, these two aspects force consumers to make compromises, as discussed in Sect. 6.1. And as the world moves further towards an interconnected society, such compromises would produce greater risks to consumers. Under current circumstances, we recommend consumers to assess relevant privacy risks carefully before making a purchase of a cleaning robot. We also anticipate more assessments and evaluations on different models of cleaning robots from a consumer-friendly perspective so that consumers could use our research results as a guideline for choice.

References

1. 262588213843476: Encryption and decryption tool for Xiaomi Mi Home's API. https://gist.github.com/socram8888/4b8949023c8e8153970764d20c935785
2. Android: Application Sandbox. https://source.android.com/security/app-sandbox
3. Android Developers: Manifest.permission. https://developer.android.com/reference/kotlin/android/Manifest.permission
4. Apple: App Store Review Guidelines - Apple Developer. https://developer.apple.com/app-store/review/guidelines/
5. Apple: Launch Services Keys. https://developer.apple.com/library/archive/documentation/General/Reference/InfoPlistKeyReference/Articles/LaunchServicesKeys.html

6. Apple: Security of runtime process in iOS and iPadOS. https://support.apple.com/guide/security/security-of-runtime-process-sec15bfe098e/web
7. ASUS: [Wireless Router] How to configure Router to use Pi-Hole? https://www.asus.com/support/FAQ/1046062/
8. bang590: JSPatch (2022). https://github.com/bang590/JSPatch
9. Benz, P.: Implementierung Und Evaluierung Eines Systems Zur Hinderniserkennung Und Kollisionsvermeidung Für Indoor-Quadrokopter. Ph.D. thesis (2013). https://www.informatik.uni-wuerzburg.de/fileadmin/10030800/user_upload/quadcopter/Abschlussarbeiten/Hinderniserkennung_Infrarot_Paul_Benz_BA.pdf
10. Bitkom: Die Zukunft der Consumer Technology – 2020, p. 61 (2020). https://www.bitkom.org/sites/default/files/2020-08/200826_ct_studie_2020_online.pdf
11. Buffalo: AirStation™ HighPower N300 Open Source DD-WRT Wireless Router. https://www.buffalotech.com/products/airstation-highpower-n300-open-source-dd-wrt-wireless-router
12. Bujlow, T., Carela-Español, V., Barlet-Ros, P.: Extended independent comparison of popular deep packet inspection (DPI) tools for traffic classification (2014)
13. Dennis, G.: Dreame Rooting. https://builder.dontvacuum.me/dreame/cmds-reset.txt
14. Dennis, G.: Vacuum Robot Overview. https://dontvacuum.me/robotinfo/#root
15. DreameTech: Dreame Bot D9 Pro Robot Vacuum and Mop (EU) User Manual. https://cdn.shopify.com/s/files/1/0302/5276/1220/files/Dreame_Bot_D9_Pro_Robot_Vacuum_and_Mop_EU_User_Manual.pdf?v=1629773800
16. DreameTech: Dreame D9 Mistral Pro — Dreame [EN]. https://dreameeurope.com/en/dreame-devices/dreame-d9-mistral-pro/#ancla
17. EUR-lex: EUR-Lex - 02016R0679-20160504, Regulation (EU) 2016/679 of the European Parliament and of the Council of 27 April 2016. https://eur-lex.europa.eu/legal-content/EN/TXT/?uri=CELEX%3A02016R0679-20160504
18. European Commission: EU: Attitudes on Data Protection and Electronic Identity in the European Union. https://joinup.ec.europa.eu/collection/eidentity-and-esignature/document/eu-attitudes-data-protection-and-electronic-identity-european-union
19. European Data Protection Supervisor: Personal Information Management System. https://edps.europa.eu/data-protection/our-work/subjects/personal-information-management-system_en
20. Fing: Fingbox. https://www.fing.com/products/fingbox
21. Fing: [Fingbox] How does Fingbox block device? https://community.fing.com/discussion/4199/fingbox-how-does-fingbox-block-device
22. Haun, M.: Einleitung als Motivation. In: Haun, M. (ed.) Handbuch Robotik. V, pp. 1–32. Springer, Heidelberg (2013). https://doi.org/10.1007/978-3-642-39858-2_1
23. Inkwire Tech: App Privacy Insights. https://apps.apple.com/us/app/app-privacy-insights/id1575583991
24. Kim Jong Cracks: Checkra1n. https://checkra.in/
25. Könings, B., Schaub, F., Weber, M.: Who, how, and why? Enhancing privacy awareness in ubiquitous computing. In: 2013 IEEE International Conference on Pervasive Computing and Communications Workshops (Percom Workshops), pp. 364–367. IEEE (2013)
26. Könings, B., Schaub, F., Weber, M., Kargl, F.: Towards territorial privacy in smart environments. In: 2010 AAAI Spring Symposium Series (2010)
27. Miller, J.F.: Supply chain attack framework and attack patterns. Technical report, MITRE Corp, Mclean, VA (2013)

28. Mitmproxy: Mitmproxy - an interactive HTTPS proxy. https://mitmproxy.org/
29. Network Working Group: Extensible Messaging and Presence Protocol (XMPP): Core. https://xmpp.org/rfcs/rfc3920.html
30. Norberg, P.A., Horne, D.R., Horne, D.A.: The privacy paradox: personal information disclosure intentions versus behaviors. J. Consum. Aff. **41**(1), 100–126 (2007). https://doi.org/10.1111/j.1745-6606.2006.00070.x
31. Okano, M.T.: IOT and industry 4.0: the industrial new revolution. In: International Conference on Management and Information Systems, vol. 25, p. 26 (2017). https://www.researchgate.net/profile/Marcelo-Okano-2/publication/319881057_IOT_and_Industry_40_The_Industrial_New_Revolution/links/59c018a5aca272aff2e20639/IOT-and-Industry-40-The-Industrial-New-Revolution.pdf
32. Pi-hole: Pi-hole®, Network-wide Ad Blocking. https://pi-hole.net/
33. Sami, S., Dai, Y., Tan, S.R.X., Roy, N., Han, J.: Spying with your robot vacuum cleaner: eavesdropping via lidar sensors. In: Proceedings of the 18th Conference on Embedded Networked Sensor Systems, SenSys 2020, pp. 354–367. Association for Computing Machinery, New York (2020). https://doi.org/10.1145/3384419.3430781
34. StatCounter: Mobile Operating System Market Share Worldwide. https://gs.statcounter.com/os-market-share/mobile/worldwide
35. Sunyaev, A.: Internet Computing: Principles of Distributed Systems and Emerging Internet-Based Technologies. Springer, Cham (2020). https://doi.org/10.1007/978-3-030-34957-8
36. unknownFalleN: Xiaomi DNS Blocklist (2022). https://github.com/unknownFalleN/xiaomi-dns-blocklist/blob/f8d3ca891e3aec4d11ffd8de23c7a7657b9a76a9/xiaomi_dns_block.lst
37. Valetudo: Valetudo. https://valetudo.cloud/
38. Vodafone: Vodafone Station - Anleitungen & Einrichtung — Vodafone-Hilfe. https://www.vodafone.de/hilfe/router/station.html
39. Weiser, M.: Some computer science issues in ubiquitous computing. Commun. ACM **36**(7), 75–84 (1993)
40. Westin, A.F.: Privacy and freedom Atheneum. New York **7**, 431–453 (1967)
41. Wireshark: Wireshark · Go Deep. https://www.wireshark.org/
42. Xiaomi: Xiaomi Home - xiaomi smarthome im App Store. https://apps.apple.com/de/app/xiaomi-home-xiaomi-smarthome/id957323480
43. Xiaomi Home: Rc4_miio/LSApplicationQueriesSchemes.txt at main · seminar-mi-robot/rc4_miio. https://github.com/seminar-mi-robot/rc4_miio
44. Yanxiu, W.: Seminar-mi-robot/rc4_miio. https://github.com/seminar-mi-robot/rc4_miio

Resource Efficient Federated Deep Learning for IoT Security Monitoring

Idris Zakariyya[1(✉)], Harsha Kalutarage[1], and M. Omar Al-Kadri[2]

[1] School of Computing, Robert Gordon University, Aberdeen, UK
{i.zakariyya,h.kalutarage}@rgu.ac.uk

[2] School of Computing and Digital Technology, Birmingham City University, Birmingham, UK
omar.alkadri@bcu.ac.uk

Abstract. Federated Learning (FL) uses a distributed Machine Learning (ML) concept to build a global model using multiple local models trained on distributed edge devices. A disadvantage of the FL paradigm is the requirement of many communication rounds before model convergence. As a result, there is a challenge for running on-device FL with resource-hungry algorithms such as Deep Neural Network (DNN), especially in the resource-constrained Internet of Things (IoT) environments for security monitoring. To address this issue, this paper proposes Resource Efficient Federated Deep Learning (REFDL) method. Our method exploits and optimizes Federated Averaging (Fed-Avg) DNN based technique to reduce computational resources consumption for IoT security monitoring. It utilizes pruning and simulated micro-batching in optimizing the Fed-Avg DNN for effective and efficient IoT attacks detection at distributed edge nodes. The performance was evaluated using various realistic IoT and non-IoT benchmark datasets on virtual and testbed environments build with GB-BXBT-2807 edge-computing-like devices. The experimental results show that the proposed method can reduce memory usage by 81% in the simulated environment of virtual workers compared to its benchmark counterpart. In the realistic testbed scenario, it saves 6% memory while reducing execution time by 15% without degrading accuracy.

Keywords: Distributed machine learning · Edge devices · Federated learning (FL) · Deep Neural Network (DNN) · Internet of Things (IoT) · Security monitoring

1 Introduction

The Internet of Things (IoT) is an ecosystem that consists of multiple intelligent devices. The Markit estimates suggest that 125 billion devices will be part of IoT by 2030 [1]. The connected IoT devices are potentially used in smart-home, smart cities, intelligent automation and cyber-physical systems. These devices

© The Author(s), under exclusive license to Springer Nature Switzerland AG 2022
W. Li et al. (Eds.): ADIoT 2022, LNCS 13745, pp. 122–142, 2022.
https://doi.org/10.1007/978-3-031-21311-3_6

used embedded systems, such as processors, sensors and communication hardware to collect and exchange data. These devices share the collected data with other connected edge devices. The shared data can improve data management and monitoring, human-machine interaction, and Artificial Intelligence (AI) analytics. However, these devices are becoming potential avenues for various cyber attacks and other cyber mafias. For instance, the massive Distributed Denial-of-Service (DDoS) attack on insecure IoT devices powered by a virus called Mirai (Linux. Gafgyt) causes a severe disaster [2]. At the same time, these IoT devices consist of low computational power, and limited memory and processors. Because of that, AI techniques developed for mainstream and other general purposes computing devices cannot be deployed on resource-constrained IoT devices. Therefore, the mechanisms to address security challenges in the IoT and cyber-physical systems need to be resource-efficient and effective, especially in Federated Learning (FL) scenario that augments data security and privacy issues.

Recent research has shown the potential applications of ML algorithms, especially Deep Neural Network (DNN), in cyber security monitoring [3]. However, IoT devices are resource-constrained and distributed in nature hence DNN-based cyber security scheme cannot be directly deployed for security monitoring in IoT environments. In addition, organizations are concerned about privacy in data sharing for training AI-based techniques in a centralised manner (e.g. data centre). In this aspect, FL [4] approach provides a promise but may not scale through IoT and cyber-physical devices because client edge devices are usually more resource-constrained in terms of storage, computational power, communication bandwidth and memory than server machines in the data centre. Therefore, training a federated DNN model consisting of millions of parameters on resource-constrained IoT devices is a challenge. To this end, we investigate the following research questions (RQs) to develop a suitable federated DNN-based method for the security monitoring of resource-constrained environments such as IoT.

- RQ1: Can existing DNNs be trained efficiently in FL settings so that the resulting model can be appropriate for IoT security monitoring in resource-constrained environments? (see Sect. 3.2)
- RQ2: Can the resulting Resource Efficient Federated DNN (REFDL) effectively and accurately detect attacks on IoT networks without accuracy degradation? (see Sect. 5)

For our experiments, we utilize a Federated Averaging (FedAvg) DNN along with eight IoT benchmark datasets to build an REFDL model. The experimental results are encouraging as the resulting REFDL shows lower memory consumption with better classification performance in simulated and real testbed federated settings against each data set used in our experiments. The federated integration of the model also helps to preserve the privacy of IoT device data during on-device model training.

The rest of the paper is organized as follows. Section 2 presents the related work. Section 3.2 describes the proposed method and the utilized FL technique, while Sect. 4 describes the evaluation process. Results and discussion can be found in Sect. 5. Finally, Sect. 6 concludes the paper with future research directions.

2 Related Work

This section presents related studies concerning deep learning for IoT security monitoring, followed by brief descriptions of FL and its applications to IoT environments.

Deep Neural Network in IoT. Significant research has been conducted on IoT security monitoring using AI techniques. Most of these methods utilized DNN [5]. Mohammad et al. [6] used DNN for IoT data analysis and network traffic classification tasks. Li et al. [7] carried out similar tasks using IoT smart cities data. Shen et al. [8] proposed compact structure-based learning with Convolutional Neural Network (CNN) for an IoT resource-constrained environment. The technique demonstrates its potentiality on the CIFAR-10 and Imagenet benchmark datasets. The lack of model assessments with IoT benchmark datasets and non-consideration of memory usage are the restrictions of their method's potentiality for deployment in a resource-constrained environment. Rock et al. [9] quantized CNN for inference on radar sensor data. Kodali et al. [10] exploit the potentiality of DNN, specifically Fully Connected Neural Network (FCNN), for classification tasks on ultra low power IoT devices. The aim is to improve the detection performance without reducing the model complexity. The lack of consideration for model complexity while selecting the FCNN architecture may restrict method feasibility. In addition, most of the stated optimization approaches considered the quantization of weights and bias parameters. Zakariyya et al. [11] proposed resource-efficient and robust DNN methods for IoT security monitoring in centralized settings. But, our proposed approach in this paper aims to reduce memory and time usage without degrading accuracy significantly in decentralized settings. The method exploits pruning, simulated micro-batching and parameter regularization to optimize the resulting model in terms of memory requirements and accuracy performance. This is useful, especially for the task of distributed learning in a resource-constrained environment.

FL in IoT Environments. McMahan et al. [12] proposed the first FedAvg FL technique that enables the training of a local model on multiple clients without sharing the client's local data to a server. This technique offers a promise in terms of model convergence with various client local data in non-independent and non-identically distributed settings. For this reason, researchers from several disciplines explored FL methods from different perspectives. In the field of IoT security monitoring, FL is gaining popularity. Preuveneers et al. [13] described FL applications for intrusion detection in IoT networks. Imteaj et al. [14] described the open research directions regarding the FL applications on resource-constrained IoT devices. Thein et al. [15] described the capability of FL in detecting attacks on industrial IoT devices. Liu et al. [16] enhance that investigation by considering raw sensor reading data. Jiang et al. [17] utilized model pruning for efficient FL training on edge devices. Bonawitz et al. [18] proposed a scalable FL framework for mobile devices to reduce communication overhead. Popoola et al. [19] used FL to detect a zero-day attack in an IoT network environment. Their implementation take the advantage of FL data privacy without considering resource limitations. Zakariyya et al. [20] used model parameters pruning and data parallelism (micro-batching) in optimizing FL to reduce memory consumption on IoT networks. However, none of the mentioned proposals considers optimizing FL to save memory and time resources using different DNN variants. We address this challenge by optimizing the federated training procedure using raw network traffic datasets from various IoT devices. Then, we proposed a REFDL method with minimal resource consumption. This method maintains state-of-the-art accuracy while reducing memory and time consumption in both simulated and real embedded devices experimental settings.

3 Methodology

To demonstrate the proof of concept, we will use Baseline Federated Deep Learning (BFDL) with FCNN and CNN model variation against some IoT and non-IoT benchmark datasets and exploit the BFDL optimization algorithm to obtain the REFDL. We demonstrate that careful optimization of the BFDL algorithm is sufficient to produce the REFDL. The efficient REFDL can detect attack activities on IoT and accurate classification with non-IoT datasets.

3.1 Baseline Federated Deep Learning (BFDL)

The BFDL utilized the classical FedAvg algorithm with an integrated DNN (FCNN/CNN) model. The in-cooperated DNN is a neural network containing deep layers of neurons representing the input data. These neurons correspond to the computing units that can transmit computational results operated with their activation function and the input. The FCNN is a sequential form of DNN that connects neurons with the corresponding weights and bias parameters. The weights and biases serve as information storage components. The baseline model of the BFDL ($\mathcal{M}_n$) in Algorithm 1 consists of network topology, activation functions and corresponding values for weights and bias. The weight and bias values settings can minimize the error function $\mathcal{E}_{\mathcal{M}_n}$ evaluated over the labelled training data $\mathcal{D}_{tr}$. This procedure can built a single master model of the FedAvg algorithm that can serve as the aggregated of the client models. The function BASE in line 1 of Algorithm 1 describes the $\mathcal{M}_n$ training using a Stochastic Gradient Descent (SGD) algorithm with backpropagation [21] in FL scenario. At each communication round, the server in function Device UPDATE of Algorithm 1 is capable of distributing a master model to each client's subsets. Each client performs iterative rounds of gradient descent weights update with their local data and returns to the server in Algorithm 3. This is determined to minimize the cost function in Eq. 1 and Eq. 2 in-order to create a global master model. Then, the execution time and memory footprints are estimated based on lines 11 and

12 of Algorithm 1. These are the records of training resource usage at the device level after the local weights update. As expressed in line 17, computed model weights are returned to the coordinating server in Algorithm 3. The server is responsible for averaging the return weight for global model aggregation. With a function that learns from $\mathcal{D}_{tr}$, the global model can appropriately map unseen samples. The resulting BFDL approach uses supervised DNN (FCNN and CNN) as a classifier, $\mathcal{M}_n$ can accept an input $\mathcal{D}_{tr}$ and outputs a probability class of vector $\hat{Y}$. The desired output $\hat{Y}$ are rounded up to the closest integer using a specified threshold value t as in Eq. 3. This output represents either the benign (1) or the attack (0) traffic instance against the IoT data or representative class for the image dataset.

Algorithm 1. Baseline BFDL training

Input: Labelled data $\mathcal{D}_{tr}$, Iteration number $\mathcal{T}$, Batch size $\mathcal{S}$
Output: Baseline model $\mathcal{M}_n$

1: **function** BASE($\mathcal{D}_{tr}[\,]$) ▷ Training baseline model
2:     **for** $i = 1$ to $\mathcal{T}$ **do**
3:         Mini-batch $B = \{(x_1, y_1), ..., (x_m, y_m)\} \subset D_{tr}$
4:         $F_p(B)$ ▷ Forward propagation with B
5:         $\mathcal{E}_i \leftarrow L$ ▷ L = Base loss
6:         B_p(B) ▷ Backward propagation
7:         **function** DEVICE UPDATE((d)) ▷ Run on device d
8:             $B_s \leftarrow$ (data P_d in batches of size B)
9:             **for** batch $b \in B_s$ **do**
10:                 $w \leftarrow$ local weights update ▷ device local weights update computation
11:                 Estimate m_i ▷ Execution memory at epoch i
12:                 Estimate t_i ▷ Execution time at epoch i
13:                 $\mathcal{M}_n$ = Trained model that estimate $\mathcal{E}_i, m_i, t_i$
14:             **end for**
15:         **end function**
16:     **end for**
17:     **return** w to server in Algorithm 3 ▷ Calls to coordinating server in Algorithm 3 for weights averaging
18:     **return** $(\mathcal{M}_n, \mathcal{E}_i, m_i, t_i)$
19: **end function**

$$J(W, b) = \frac{1}{m}\sum_{i=1}^{m} L(\hat{Y}^i, Y^i) \tag{1}$$

$$L(\hat{Y}^i, Y^i) = -(Y\log\hat{Y} + (1 - Y)\log(1 - \hat{Y})) \tag{2}$$

$$Output = \begin{cases} 0 \text{ if } \hat{Y} \leq t \\ 1 \text{ if } \hat{Y} > t \end{cases} \tag{3}$$

3.2 Resource Efficient Federated Deep Learning (REFDL)

As mentioned above, training a resource-efficient DNN model for FL task can be a challenging task, especially in IoT security monitoring [22]. Because of the FL communication rounds and DNN model parameters requirements in designing and building the desirable architecture. The complexity of such an approach increases with multidimensional datasets.

Algorithm 2. Proposed method to obtain REFDL

Input: Penalty term λ, ($\mathcal{D}_{tr}, \mathcal{T}, B, L$, in Algorithm 1)
Output: Efficient model $\mathcal{M}_e$

1: **function** EFFICIENT($\mathcal{D}_{tr}[\,]$)
2: **for** $j = 1$ to $\mathcal{T}$; **do**
3: Micro-batch $M = \{(x_1, y_1), ..., (x_m, y_m)\} \subset B$
4: $F_p(M)$ ▷ Forward propagation with M
5: $\mathcal{E}_t = L$ ▷ Initialized loss
6: Estimate m_t, t_t Initialized memory and time based on $\mathcal{E}_t$
7: $\mathcal{E}_j \leftarrow \mathcal{E}_t + \lambda \sum_{j=1}^{W} \frac{(w_j^2/w_0^2)}{(1+w_j^2/w_0^2)}$
8: B_p(M) ▷ Backward propagation with M
9: **function** DEVICE UPDATE((d)) ▷ Run on device d
10: $M_s \leftarrow$ (data P_d in batches of size M)
11: **for** batch $b \in M_s$ **do**
12: $w \leftarrow$ local weights update ▷ device local weights update computation
13: **if** $(\mathcal{E}_j \leq \mathcal{E}_t)$ **then**
14: $\lambda = \lambda + \triangle\lambda$
15: Estimate m_j ▷ Execution memory at epoch j
16: Estimate t_j ▷ Execution time at epoch j
17: **if** $((m_j < m_t) \wedge (t_j < t_t))$ **then**
18: $m_{tr} = m_j$ ▷ $m_t r$ = Efficient memory
19: $t_{tr} = t_j$ ▷ t_{tr} = Efficient time
20: $\mathcal{M}_e$ = Trained model that estimate $\mathcal{E}_j, m_{tr}, t_{tr}$
21: **end if**
22: **end if**
23: **end for**
24: **end function**
25: **end for**
26: **return** w to server in Algorithm 3 ▷ Calls to coordinating server in Algorithm 3 for weights averaging
27: **return** $(\mathcal{M}_e, \mathcal{E}_j, m_{tr}, t_{tr})$
28: **end function**

To this end, we utilize the baseline model in BFDL to produce its resource-efficient counterparts (REFDL). The training procedure described in Algorithm 2 optimizes a function using $\mathcal{D}_{tr}$ in the FL scenario to obtain the efficient M_e corresponding to the REFDL model. As described in line 3 in Algorithm 2, the optimization procedure utilized micro-batching [23], which is suitable for breaking a large amount of data into smaller batches for efficient on-device model training. Unlike the mini-batch, the micro-batching is particularly suitable for most datasets, especially the IoT ones. To reduce network complexity, we used a penalty [24] (weight elimination) technique with a threshold parameter w_0 as shown in regularized Eq. 4. This is a requirement to discover those sets of relevant weights in the network for efficient local weight updates. In particular, to determine the significant and insignificant large weights of the baseline model. Weights greater than w_0 that yield a complexity cost closer to 1 require a regularization using the penalty parameter λ. However, we do not predefined the numbers of weights to eliminate, the Algorithm 2 itself will decide this number based on the given DNN architecture and the various other constraints. The regularization considers a scenario where the initialized model produces a higher error value $\mathcal{E}_t$ as in line 7. For better performance, we utilized the set of parameters to produce a lower error value $\mathcal{E}_j$. After this stage, in lines 15 and 16, the estimated computational memory footprints and execution time are compared with that of the initialized values in line 6 to return the minimal memory constraint produced by the client device model. Device models with minimal resource consumption are returned to the coordinating server in Algorithm 3 together with their weights for model averaging. Then, the coordinating server can update the client model weights in a federated setting and performs weight averaging while returning the updated averaged weights for model aggregation. This process can reduce the client's communication time and computational complexity while building the aggregate model of REFDL. The memory and processor savings for each client device at each federated round and accumulating all these savings can lead to significant savings when the model is converged.

$$R = \lambda \sum_{j=1}^{W} \frac{(w_j^2/w_0^2)}{(1 + w_j^2/w_0^2)} \tag{4}$$

Algorithm 3. Coordination Procedure for Algorithms 1 and 2

Server Executes:

1: **function** SERVER WEIGHTS UPDATE
2: initialize weight w;
3: **while** $t \leq n$ **do** ▷ n federated round
4: $R \leftarrow$ random set of $max(C.K, 1)$ ▷ $C.K$ fraction of clients K
5: **for** $k \in R$ in parallel **do** ▷ k client index
6: Weight device update ▷ Federated model weight update for Algorithms 1 or 2
7: **end for**
8: Averaged weights update
9: **end while**
10: **return** Averaged updated weights
11: **end function**

4 Evaluation

This section describes the evaluation criteria of the BFDL and REFDL methods. It also presents the datasets used in the evaluation of the proposed approach.

4.1 Utilized Datasets

The N-BaIoT dataset contains various realistic data samples from nine commercial IoT devices that collectively represent multitudes of botnet and benign network traffic flows [25]. Each device is either infected by a variety of BASHLITE or Mirai attacks, with some regular instances. We randomly consider eight devices subsets with the most IoT specification. These devices are a (i) Danmini Doorbell, (ii) Ecoobee Thermostat, (iii) Ennio Doorbell, (iv) Provision PT-737E, (v) Provision PT-838, (vi) Samsung SNH-1011-N, (vii) SimpleHome XCS-1002-WHT, and (ix) SimpleHome XCS-1003-WHT. Each device consists of sufficient records of variational attacks such as ack, syn, scan, junk, tcp, udp, udpplain, combo and regular instances containing a numeric representation of traffic flows with 115 features vector. For this reason, the N-BaIoT dataset serves as a benchmark for the proposal of a device-centric IoT security monitoring mechanism. We utilized mentioned commercial devices subsets data of the N-BaIoT for federated training and testing of BFDL and REFDL models.

The WUSTL dataset consists of multiple flows of traffic from an emulated SCADA system [26]. This dataset can be appropriate for investigating the feasibility of AI algorithms for security monitoring purposes. The raw data consists of 7,037,983 numeric data samples. As a result, we consider the distribution of 471,545 attacks and 6,566,438 normal instances to evaluate our method.

4.2 Virtual Workers Experimental Setup and Implementation

We used Python 3.76 on a desktop computer with Intel Xeon E5-2695(4 core) CPUs running at 2.10 GHz with 16.0 GB installed memory to build each technique. For profiling memory consumption, we utilized the integrated memory usage [27]. We utilized PyTorch version 1.4.0 [28] and PySyft version 0.2.9 [29] frameworks for the virtual on-device training. Pysyft framework simplifies the creation of virtual workers. We utilized these virtual workers to simulate the FL scenario for the BFDL and REFDL. These workers emulate real virtual machines and can run as a separate process within the same python program with their dataset. Our federation training procedure considered four clients' virtual workers and a coordinating server worker receiving the computational updates from each virtual client worker model. Each federated client model consists of an input layer, four hidden layers and an output layer. The topology selection against each dataset utilized [30] to minimize operations and improve the performance metrics. The experimental settings considered are appropriate for binary classification as returned by the [30] parameter tuning technique. The overall architectural settings remain identical for evaluating the BFDL and proposed REFDL technique. Table 1 presents the utilized model architecture of each

FL technique against each dataset. Regarding the Wustl dataset, the selected topology was returned based on [30] tuning technique.

Table 1. Architecture and distribution of normal and attack for each device data.

Device	Normal	Attack	Inputs	Outputs	Architecture
Danmini Doorbell	49,548	968,750	115	1	83-128-128-83
Ecobee Thermostat	13,113	822,763	115	1	83-128-128-83
Ennio Doorbell	39,100	316,400	115	1	83-128-128-83
Provision PT-737E	62,154	766,106	115	1	83-128-128-83
Provision PT-838	98,514	729,862	115	1	83-128-128-83
Samsung SNH-1011-N	52,150	323,072	115	1	83-128-128-83
SimpleHome XCS-1002-WHT	46,585	816,471	115	1	83-128-128-83
SimpleHome XCS-1003-WHT	19,528	831,298	115	1	83-128-128-83
Wustl	6,566,438	471,545	6	1	26-128-128-26

For the baseline and optimized model training procedure, $lr = 0.001$ was utilized. We used 0.01 values for both λ, $\triangle\lambda$ and threshold w_0 [31] with 4 micro-batches to build the model of REFDL method. The activation function considered in the fully connected layers is relu [32] with sigmoid in the output layer. Both BFDL and REFDL use an SGD optimizer appropriate for running FedAvg training. Each federated model was trained in 128 batches within four epochs in 30 worker's communications rounds for optimum convergences. After completing the client's model training, average weight values are sent to the coordinating worker. This worker aggregates those weights to update the global model. Codes for this implementation are made publicly accessible for exploration and reproduction purposes [33].

4.3 Testbed Experimental Setup and Implementation

To test the efficient federated communication of the REFDL against BFDL in a testbed setting, we utilized the PySyft version 0.2.9 [29] python framework over a network with a client and server-class connected via a WebSocket (WS). Since PyTorch is a potential library for PySyft, we utilize it to build an edge computing FL training scenario for resource-constrained devices. The environmental settings mimic the client's server communication scenario in a distributed manner. In this context, it can support the building of simulated and realistic testbed settings. In the network realistic testbed settings, we considered 4 Gigabyte Brix (GB-BXBT-2807) with a laptop (see Fig. 1). The personal laptop represents the coordinating server in a wireless network to emulate low-frequency connections. The server is responsible for model weights aggregation and distribution to clients. The client's devices in Algorithms 1 and 2 are responsible for local model training using the server model weights on the client's dataset and returning client weights to the server. Therefore, the communication workload is higher at the client-side containing the edge devices than the server machine. The installed Operating System (OS) in GB-BXBT-2807 clients is Ubuntu version 20.04.4 LTS. Each client contains an installation for the PySyft framework and its dependencies. Federated network testbed implementations codes are publicly accessible [34].

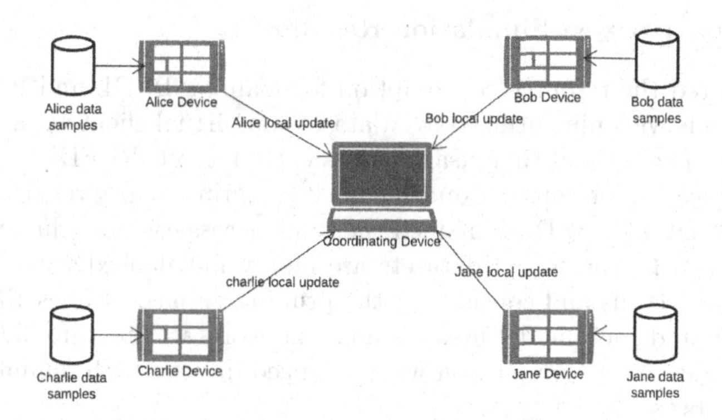

Fig. 1. BFDL and REFDL model training testbed with gigabyte devices.

For evaluating the simulated runtime and real execution time of BFDL and REFDL, experiments with four workers with their distributed training data (Alice, Bob, Charlie and Jane as shown in Fig. 1) were performed. A federated communication round of 50 is used, with two epoch iterations, within a 64 mini-batch size as returned by the optimized tuning procedure. The test batch sample size selection is 1000 with $lr = 0.01$ for effective FedAvg SGD training. The utilized real-time models for each federated client of both Algorithm 1 and 2 contain an input layer and four identical hidden layers (128-128-128-128) with an output layer or layers as the case may be. The chosen architecture can support effective and efficient model convergence. To test the REFDL effectiveness and generalizability, we considered the CNN DNN variant in realistic settings with clients utilizing the MNIST image dataset [35]. The CNN architecture contains two convolutional layers (Conv-2). The first 2D convolutional layer requires one input to output 20 convolutional features using a 5 square kernel (1, 20, 5, 1). The second 2D convolutional layer requires 20 input layers to output 50 convolutional features using a 3 square kernel (20, 50, 5, 1). The architecture in the first real-time layer is (800 (4*4*50), 128) with (128, 10) in the second real-time layer. Max-Pool in 2d was run over the input image without a dropout utilization. The fully connected hidden layers in the convolutional are similar to the version described in Table 1.

5 Results and Discussion

This section discusses the experimental results. It details the evaluation comparison of the optimized REFDL and baseline BFDL FedAvg models in simulation and testbed settings across datasets.

5.1 Virtual Workers Simulation Results

We investigated the resource consumption for training BFDL and REFDL federated methods with nine utilized IoT datasets on virtual client workers. Table 2 presents the memory and time usage across each dataset. REFDL training procedure produces lower runtime and memory footprints. However, the accuracy for both REFDL and BFDL remained the same across each benchmark dataset. The reason can be the tested datasets are highly imbalanced with large number of testing records and considering the pruning applied in lines 13 and 14 of Algorithm 2, and each model uses a similar network architecture [36]. Refer to Table 4 and Fig. 10 for comparison with balanced dataset with minimal number of testing records.

Table 2. Federated model training memory consumption between REFDL and BFDL.

Dataset	Model	Memory MB	Time minutes	Test acc %
Danmini Doorbell	BFDL	3.783	0.099	95.11
	REFDL	0.857	0.081	95.11
Ecobee Thermostat	BFDL	3.732	0.091	93.36
	REFDL	0.815	0.071	93.36
Ennio Doorbell	BFDL	4.147	0.090	88.94
	REFDL	0.805	0.074	88.94
Provision PT-737E	BFDL	3.463	0.092	92.52
	REFDL	0.853	0.077	92.52
Provision PT-838	BFDL	3.423	0.085	88.07
	REFDL	0.814	0.074	88.07
Samsung SNH-1011-N	BFDL	3.783	0.099	86.06
	REFDL	0.858	0.081	86.06
SimpleHome XCS-1002	BFDL	3.494	0.090	94.65
	REFDL	0.816	0.072	94.65
SimpleHome XCS-1003	BFDL	3.914	0.085	97.73
	REFDL	0.801	0.071	97.73
Wustl	BFDL	3.002	0.095	94.26
	REFDL	0.816	0.076	94.26

Figures 2 and 3 show the percentage of memory and time reduction by REFDL as reflected in Table 2. The results demonstrate a significant percentage of memory saving across each dataset. Regarding client processing runtime, REFDL is more efficient. It indicates less complexity, faster learning capability and effective performance behaviour over BFDL. These resources minimization make it a better choice for IoT security monitoring. Especially for the on-device learning across various distributed resource-constrained edge devices.

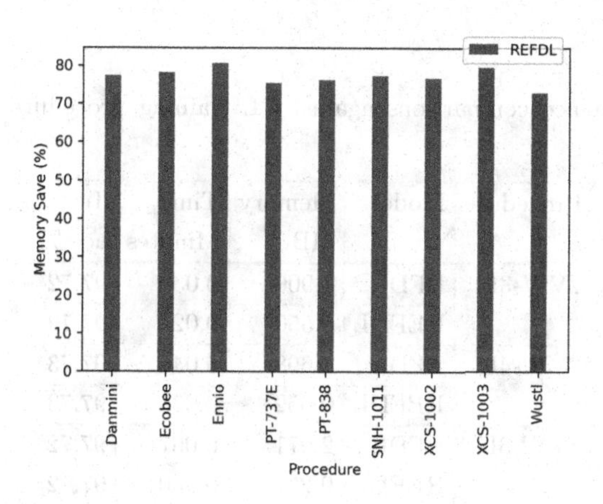

Fig. 2. REFDL federated model training memory resources saved against datasets.

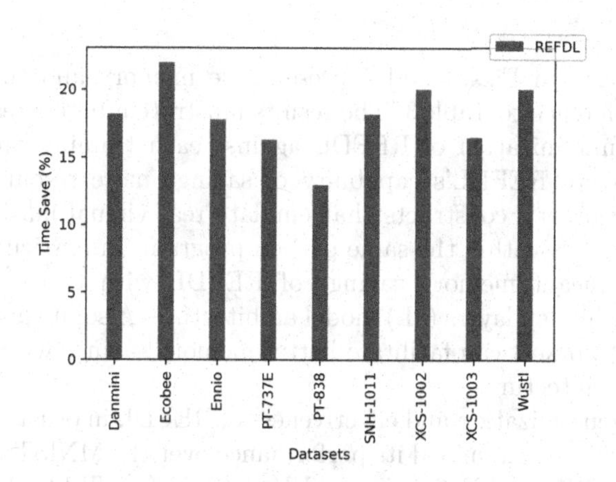

Fig. 3. REFDL federated model training time resources saved against datasets.

The results in Table 3 are for the implemented BFDL method and its optimized counterpart REFDL against training procedures. It compared the training time, memory requirements and accuracy against model hidden layers (L) and virtual workers (VW) variations. As presented, the REFDL requires lower memory and time as tested with the XCS-1003 dataset. Both BFDL and REFDL federated models produce slightly better accuracy with four hidden layers (4L). As a result, the increment of hidden layers influences effective federated learning in distributed settings. However, the clients with higher computational power can influence on the global model significantly than the clients with lower computational resources. The class imbalance across clients can also influence the accuracy of the global model. It would be interesting to investigates these limitations in future work.

Table 3. Performance comparisons against FL training procedure on SimpleHome XCS-1003 dataset.

Procedure	Model	Memory MB	Time Minutes	Test acc %
2VW-3L	BFDL	1.906	0.038	97.72
	REFDL	0.550	0.027	97.72
2VW-4L	BFDL	2.698	0.046	97.73
	REFDL	0.052	0.036	97.73
4VW-3L	BFDL	2.971	0.067	97.72
	REFDL	0.294	0.060	97.72
4VW-4L	BFDL	3.914	0.085	97.73
	REFDL	0.801	0.071	97.73

The illustration in Figs. 4 and 5 present the memory and time savings of REFDL with reference to Table 3. The results illustrate a better resource (memory and time) minimization of REFDL against each training procedure. The results demonstrate REFDL's capability of savings more resources using the PySyft virtual worker's constructs that emulate real virtual machines and run as a separate process within the same python program. In particular, it demonstrates the significant memory savings of REFDL with two virtual workers (2VW) and four hidden layers (4L) model architecture. Also, it shows that increments of virtual workers can facilitate better memory savings with three hidden layer network architecture.

To test the generalization and effectiveness of REFDL in other domains other than cybersecurity, we examined its performance over the MNIST image dataset with integrated CNN and FCNN in the FL scenario (see Table 4). This is good to assess the method's performance over non-IoT datasets so that it can be a generic solution for on-device learning. In particular, to exploit the resource-saving capability of CNN that offers a promise in image classification. In this aspect, we utilized the PySyft WS (network) simulated workers and examined the performance of the BFDL and REFDL techniques in each federated training. This is to assess REFDL performance using a simulated network with a client

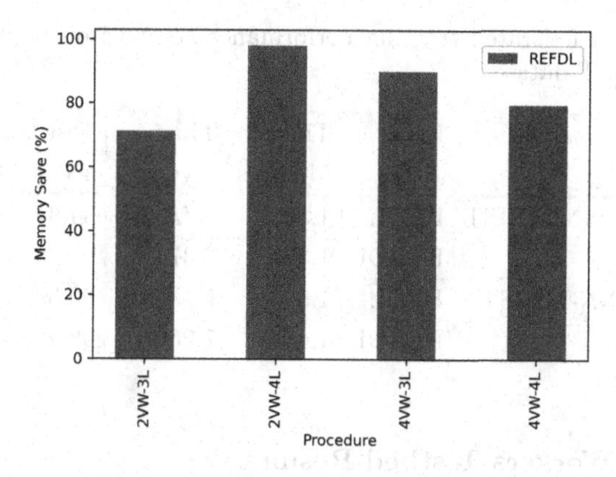

Fig. 4. REFDL federated model training memory resources saved with XCS-1003 dataset.

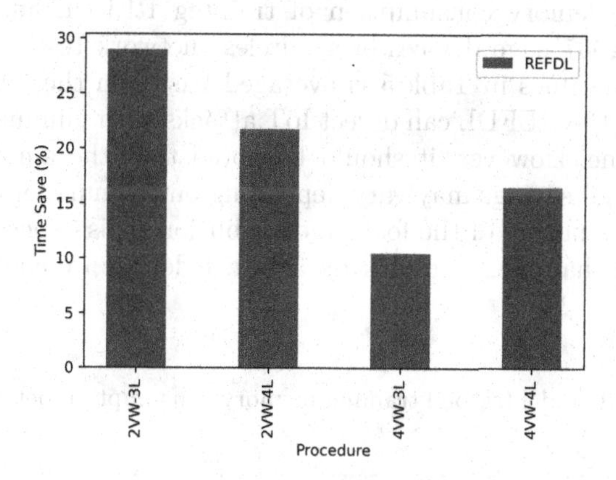

Fig. 5. REFDL federated model training time resources saved with XCS-1003 dataset.

and server scenario running on the same machine, not like PySft virtual workers counterparts that run as construct within the same python program. With each DNN (CNN and FCNN) variant, the REFDL demonstrates better accuracy than its BFDL counterparts. The reason for better performance with the MNIST dataset can be because the distribution of the dataset is highly balanced for both the training and testing cases. In addition, it produces lower training execution time. These results show the important of regularization [37,38] on accuracy against DNN variation. This attracts further investigation in realistic settings.

Table 4. Simulated federated training performance comparison between BFDL and REFDL with MNIST dataset.

Procedure	FL	Time Minutes	Time save (%)	Test set acc %
FCNN-MNIST	BFDL	1.393	N/A	34.64
	REFDL	1.346	3.374	91.03
CNN-MNIST	BFDL	1.583	N/A	90.59
	REFDL	1.457	7.960	98.28

5.2 Network Workers Testbed Results

With Ennio Doorbell and Samsung SNH randomly selected IoT datasets, we investigated the memory consumption of training REFDL and BFDL across four GB-BXBT-2807 edge devices over wireless network testbed settings. The reported memory values in Table 5 is averaged based on the four devices. The results show that the REFDL can detect IoT attacks with minimal memory than BFDL in real-time. However, it should be noted that the amount of memory and CPU (runtime) savings may vary depending on the number of clients in the federation and the nature of the feature distribution. It is expected that a large number of clients handling complex data in a federation could lead to higher savings.

Table 5. Federated model testbed training memory consumption between REFDL and BFDL.

Dataset	Model	Memory MB	Memory Save %	Test acc %
Ennio Doorbell	BFDL	33.965	N/A	89.00
	REFDL	31.981	5.84	89.00
Samsung SNH	BFDL	32.519	N/A	86.10
	REFDL	30.550	6.05	86.10

In Fig. 6, we present averaged estimated convergence real-time for training BFDL and REFDL on GB-BXBT-2807 testbed against the Ennio Doorbell and Samsung SNH IoT datasets. The REFDL requires less real-time than BFDL in detecting IoT attacks. The results demonstrate the effectiveness of REFDL in saving computational resources in resource-constrained environments. As illustrated in Fig. 7, the savings can be better with many decentralized edge devices.

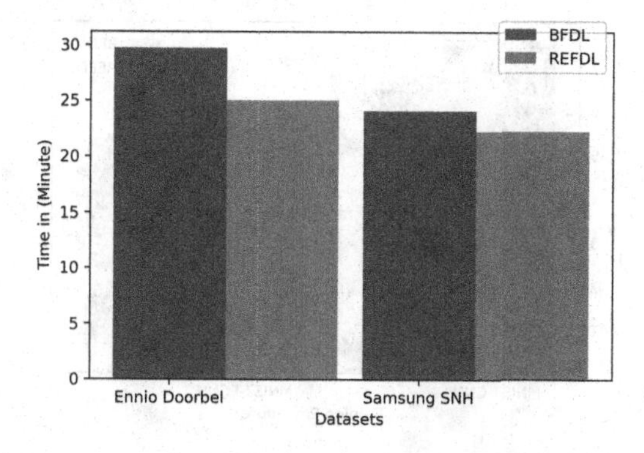

Fig. 6. Federated model training execution time of REFDL and BFDL against datasets.

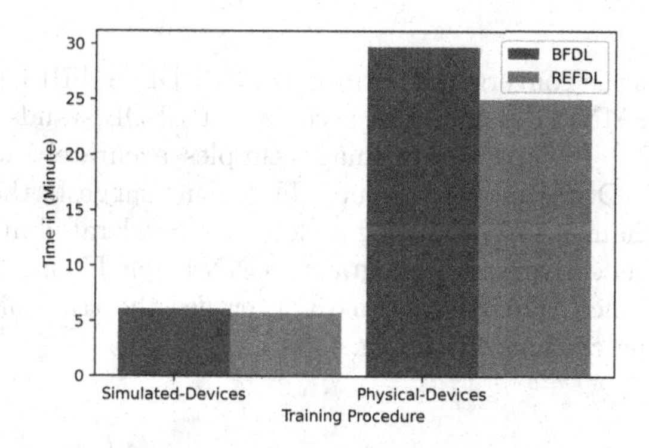

Fig. 7. REFDL and BFDL training execution time in a simulated and realistic network environment against Ennio dataset.

The result demonstrates the savings advantage of REFDL in realistic network settings over the simulated virtual WS connections counterparts. It indicates REFDL's capability in saving more resources in a resource-limited environment with multiple client devices.

We examined the real-time savings of the REFDL over BFDL against the MNIST dataset. Figure 8 show that REFDL is more efficient than BFDL across each training procedure. The MNIST-CNN federated training procedure is more computationally expensive than the MNIST-FCNN. In this context, the FCNN DNN variant of REFDL can be an appropriate choice for on-device learning if savings resources are the target objectives. In that case, REFDL stands more suitable method for deployment in an IoT resource environment.

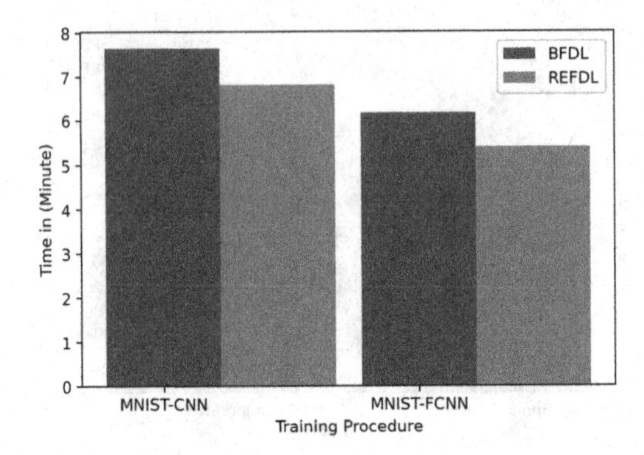

Fig. 8. Federated model training execution time usage between REFDL and BFDL with mnist dataset.

Figure 9 shows the convergence accuracy of REFDL and BFDL against DNN variants with the MNIST dataset. In each case, REFDL stands to be a better model than BFDL. It can classify image samples accurately with integrated CNN and FCNN (DNN) model variants. The result suggests the advantage of optimization mechanisms in producing a global deep federated model. It further demonstrates the effectiveness of integrating CNN in the FL method to improve accuracy performance. This is good as it leverages the tradeoff between each DNN model during on-device learning.

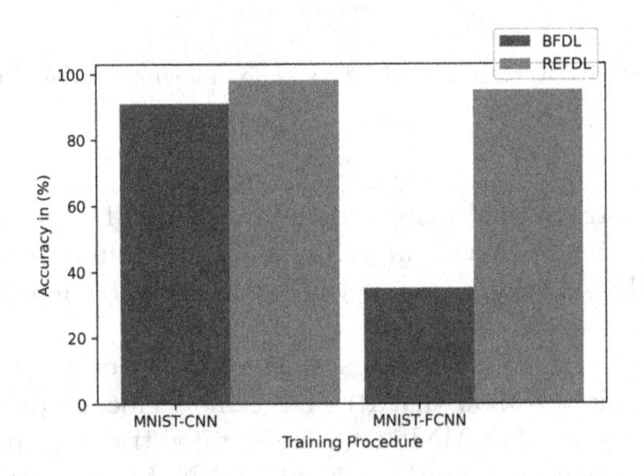

Fig. 9. Federated model accuracy comparison between REFDL and BFDL with mnist dataset.

To test the effectiveness and faster learning of REFDL on GB-BXBT-2807 testbed federated settings, we vary the epoch iterations using the FCNN-MNIST procedure. In that context, we can assess the performance of each federated method in real-time. As shown in Fig. 10, the REFDL can achieve a better accuracy even with one local epoch and 50 communication round. This trends of providing higher accuracy remain stable across each epoch iteration. The result demonstrates REFDL appropriateness and faster learning capability across edge devices, especially with the integrated FCNN model. REFDL minimum number of epoch requirements is advantageous. Especially in an environmental setting such as IoT with inherited limited memory resources.

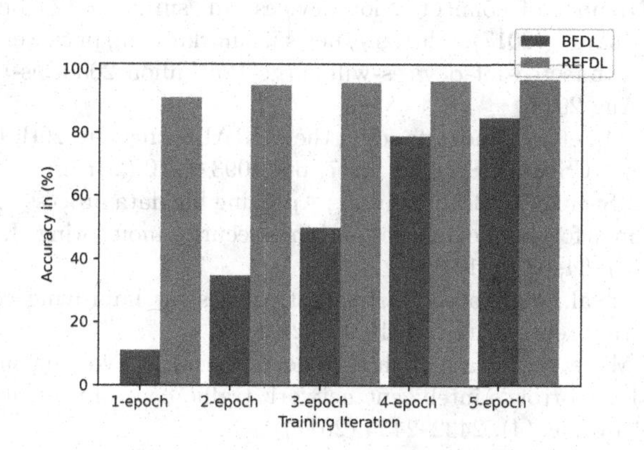

Fig. 10. Federated model accuracy performance with epochs between REFDL and BFDL against mnist dataset.

6 Conclusion

As FL uses a distributed ML to enable on-device learning in decentralized edge devices over a network with the support of data privacy across multiple clients, this paper investigated the feasibility of running FL training in resource-constrained environments such as IoT. In particular, to develop feasible and effective security solutions for IoT devices. In this paper, we utilized FedAvg (BFDL algorithm) with carefully selected model optimization techniques to produce an effective and resource-efficient REFDL federated model. The experiments evaluation with eight IoT datasets and one image dataset in simulated and GB-BXBT-2807 realistic testbed settings demonstrate the effectiveness, low complexity and efficient nature of REFDL. It detects IoT attacks accurately using minimal resources than its counterparts. Also, it can perform better in classifying image samples with fully connected and convolutional deep neural network models in a federated training scenario. In addition, REFDL requires fewer epochs to produce a more accurate FL model than its counterparts. These motivational results attract further investigation for utilizing more computational networks nodes/client devices at deployment, particularly over-wired and wireless settings using our testbed. In addition, we plan to investigate the resilient capability of the REFDL to enhance its security robustness against adversarial attacks in a realistic network setting with various connected edge devices other than the ones considered in this paper. This can enable us to examine the resource efficiency and security monitoring performance of our proposed method capability and potentiality across multiple decentralized edge devices.

Acknowledgment. This work was supported by the Petroleum Technology Development Fund (PTDF), Nigeria.

References

1. Howell, J.: Number of connected iot devices will surge to 125 billion by 2030, ihs markit says (2017). https://news.ihsmarkit.com/prviewer/release_only/slug/number-connected-iot-devices-will-surge-125-billion-2030-ihs-markit-says. Accessed 17 Aug 2018
2. Antonakakis, M., et al.: Understanding the MIRAI botnet. In: 26th USENIX security Symposium USENIX Security 2017, pp. 1093–1110 (2017)
3. Kotenko, I.V., Saenko, I., Branitskiy, A.: Applying big data processing and machine learning methods for mobile internet of things security monitoring. J. Internet Serv. Inf. Secur. **8**(3), 54–63 (2018)
4. Konečný, J., et al.: Federated learning: strategies for improving communication efficiency. arXiv preprint arXiv:1610.05492 (2016)
5. Fadlullah, Z.M., et al.: State-of-the-art deep learning: Evolving machine intelligence toward tomorrow's intelligent network traffic control systems. IEEE Commun. Surv. Tutor. **19**(4), 2432–2455 (2017)
6. Mohammadi, M., Al-Fuqaha, A., Sorour, S., Guizani, M.: Deep learning for IoT big data and streaming analytics: a survey. IEEE Commun. Surv. Tutor. **20**(4), 2923–2960 (2018)
7. Li, X., Liu, H., Wang, W., Zheng, Y., Lv, H., Lv, Z.: Big data analysis of the internet of things in the digital twins of smart city based on deep learning. Futur. Gener. Comput. Syst. **128**, 167–177 (2022)
8. Shen, S., Li, R., Zhao, Z., Liu, Q., Liang, J., Zhang, H.: Efficient deep structure learning for resource-limited IoT devices. In: GLOBECOM 2020–2020 IEEE Global Communications Conference. IEEE, pp. 1–6 (2020)
9. Rock, J., Roth, W., Toth, M., Meissner, P., Pernkopf, F.: Resource-efficient deep neural networks for automotive radar interference mitigation. IEEE J. Select. Topi. Signal Process. **15**(4), 927–940 (2021)
10. Kodali, S., Hansen, P., Mulholland, N., Whatmough, P., Brooks, D., Wei, G.-Y.: Applications of deep neural networks for ultra low power IoT. In: 2017 IEEE International Conference on Computer Design (ICCD), pp. 589–592. IEEE (2017)
11. Zakariyya, I., Kalutarage, H., Al-Kadri, M.O.: Robust, effective and resource efficient deep neural network for intrusion detection in IoT networks. In: Proceedings of the 8th ACM on Cyber-Physical System Security Workshop, pp. 41–51 (2022)
12. McMahan, B., Moore, E., Ramage, D., Hampson, S., y Arcas, B.A.: Communication-efficient learning of deep networks from decentralized data. In: Artificial Intelligence and Statistics, pp. 1273–1282. PMLR (2007)
13. Preuveneers, D., Rimmer, V., Tsingenopoulos, I., Spooren, J., Joosen, W., Ilie-Zudor, E.: Chained anomaly detection models for federated learning: an intrusion detection case study. Appl. Sci. **8**(12), 2663 (2018)
14. Imteaj, A., Thakker, U., Wang, S., Li, J., Amini, S.E.:A survey on federated learning for resource-constrained IoT devices. IEEE Internet of Things J. **9**, 1–24 (2021)
15. Nguyen, T.D., Marchal, S., Miettinen, M., Fereidooni, H., Asokan, H., Sadeghi, A.-R.: Dïot: A federated self-learning anomaly detection system for IoT. In: 2019 IEEE 39th International Conference on Distributed Computing Systems (ICDCS), pp. 756–767. IEEE (2019)

16. Liu, Y., et al.: Communication-efficient federated learning for anomaly detection in industrial internet of things. In: GLOBECOM 2020–2020 IEEE Global Communications Conference, pp. 1–6 IEEE (2020)
17. Jiang, Y., et al.: Model pruning enables efficient federated learning on edge devices. arXiv preprint arXiv:1909.12326 (2019)
18. Bonawitz, K., et al.: Towards federated learning at scale: system design. arXiv preprint arXiv:1902.01046 (2019)
19. Popoola, S.I., Ande, R., Adebisi, B., Gui, G., Hammoudeh, M., Jogunola, O.: Federated deep learning for zero-day botnet attack detection in IoT edge devices. IEEE Internet of Things J. **9** (2021)
20. Zakariyya, I., Kalutarage, H., Al-Kadri, M.O.: Memory efficient federated deep learning for intrusion detection in IoT networks. In: CEUR Workshop Proceedings (2021)
21. Chauvin, Y., Rumelhart, D.E.: Backpropagation: Theory, Architectures, and Applications. Psychology Press, Hove (2013)
22. Abiodun, O.I., Jantan, A., Omolara, A.E., Dada, K.V., Mohamed, N.A., Arshad, H.: State-of-the-art in artificial neural network applications: A survey. Heliyon **4**(11), e00938 (2018)
23. Huang, Y., et al.: Gpipe: efficient training of giant neural networks using pipeline parallelism. Adv. Neural. Inf. Process. Syst. **32**, 103–112 (2019)
24. Han, S., Pool, J., Tran, J., Dally, W.J.: Learning both weights and connections for efficient neural networks. arXiv preprint arXiv:1506.02626 (2015)
25. Meidan, Y., et al.: N-Baiot-network-based detection of IoT botnet attacks using deep autoencoders. IEEE Perv. Comput. **17**(3), 12–22 (2018)
26. Teixeira, M.A., Salman, T., Zolanvari, M., Jain, R., Meskin, N., Samaka, M.: Scada system testbed for cybersecurity research using machine learning approach. Fut. Internet **10**(8), 76 (2018)
27. Pedregosa, F., Gervais, P.: Memory Profiler (python). Python Software Foundation, https://pypi.org/project/memory-profiler/ Accessed 25 Mar 2019
28. Paszke, A., et al.: Pytorch: an imperative style, high-performance deep learning library. Adv. Neural. Inf. Process. Syst. **32**, 8026–8037 (2019)
29. Ryffel, T., et al.: A generic framework for privacy preserving deep learning. arXiv preprint arXiv:1811.04017 (2018)
30. Komer, B., Bergstra, J., Eliasmith, C.: Hyperopt-Sklearn. In: Hutter, F., Kotthoff, L., Vanschoren, J. (eds.) Automated Machine Learning. TSSCML, pp. 97–111. Springer, Cham (2019). https://doi.org/10.1007/978-3-030-05318-5_5
31. Bosman, A., Engelbrecht, A., Helbig, M.: Fitness landscape analysis of weight-elimination neural networks. Neural Process. Lett. **48**(1), 353–373 (2018)
32. Ide, H., Kurita, T.: Improvement of learning for CNN with RELU activation by sparse regularization. In: 2017 International Joint Conference on Neural Networks (IJCNN), pp. 2684–691 IEEE (2017)
33. Zakariyya, I.: Resource efficient federated algorithm with virtual workers (2022) https://github.com/izakariyya/sim-virtual-fed-dnn
34. Zakariyya, I.:Resource efficient federated algorithm with realistic workers (2022). https://github.com/izakariyya/testbd-fl-iot
35. Deng, L.: The MNIST database of handwritten digit images for machine learning research [best of the web]. IEEE Signal Process. Mag. **29**(6), 141–142 (2012)

36. García, V., Mollineda, R.A., Sánchez, J.S.: Theoretical analysis of a performance measure for imbalanced data. In : 2010 20th International Conference on Pattern Recognition, pp. 617–620. IEEE (2010)
37. Krueger, D., Memisevic, R.: Regularizing RNNs by stabilizing activations. arXiv preprint arXiv:1511.08400 (2015)
38. Lever, J., Krzywinski, M., Altman, N.: Points of significance: Regularization. Nat. Methods **13**(10), 803–805 (2016)

Man-in-the-OBD: A Modular, Protocol Agnostic Firewall for Automotive Dongles to Enhance Privacy and Security

Felix Klement[1(✉)], Henrich C. Pöhls[2], and Stefan Katzenbeisser[1]

[1] Chair of Computer Engineering, University of Passau, Passau, Germany
{felix.klement,stefan.katzenbeisser}@uni-passau.de
[2] Chair of IT Security, University of Passau, Passau, Germany
hp@sec.uni-passau.de

Abstract. Third-party dongles for cars, e.g. from insurance companies, can extract sensitive data and even send commands to the car via the standardized OBD-II interface. Due to the lack of message authentication mechanisms, this leads to major security vulnerabilities for example regarding the connection with malicious devices. Therefore, we apply a modular, protocol-independent firewall approach by placing a man-in-the-middle between the third-party dongle and the car's OBD-II interface. With this privileged network position, we demonstrate how the data flow accessible through the OBD-II interface can be modified or restricted. We can modify the messages' contents or delay the arrival of messages by using our fine-granular configurable rewriting rules, specifically designed to work protocol agnostic. We have implemented our modular approach for a configurable firewall at the OBD-II interface and successfully tested it against third-party dongles available on the market. Thus, our approach enables a security layer to enhance automotive privacy and security of dongle users, which is of high relevance due to missing message authentications on the level of the electronic control units.

Keywords: Network security · Information flow control · Vehicular security · CAN · OBD-II

1 Introduction

Today's cars, self-driving or not, can exchange information with the outside over a standardised interface: the On-Board-Diagnosis (OBD) interface, which exists since 1988. OBD-II is a vehicle diagnostic system, which makes important electronic control units (ECU) of the car and their data accessible. This allows reading the current speed, rpm and other information from the car. In 1996 the

The authors acknowledge the financial support by the Federal Ministry of Education and Research of Germany in the program of "Souverän. Digital. Vernetzt." Joint project 6G-RIC, project identification number: 16KISK034.

© The Author(s), under exclusive license to Springer Nature Switzerland AG 2022
W. Li et al. (Eds.): ADIoT 2022, LNCS 13745, pp. 143–164, 2022.
https://doi.org/10.1007/978-3-031-21311-3_7

USA made it mandatory for every new car that is sold to have this interface. According to Regulation (EC) No. 715/2007, all new passenger car registrations in the EU since 2001 for gasoline engines and since 2004 also for diesel engines must be equipped with an OBD interface. This means that nowadays a huge amount of cars are equipped with this On-Board-Diagnosis system. However, this also means that the driver is strongly encouraged or forced to connect external devices (dongles) to this port first, so that they can then read out the car's data via the OBD interface. In this paper we show that the OBD interface lacks basic data authentication mechanisms making it possible to place our firewall as a man-in-the-middle. In the remainder of this paper we will use the term *dongle* to refer to the third-party device that will be connected to the car over the OBD-II interface. Dongles might offer all functionality on their own (standalone) or consisting of a hardware connector and a mobile phone connected via bluetooth.

As mentioned above, the On-Board-Diagnostic interface allows for interaction with a variety of ECUs and to obtain valuable data for an overall insight into the current state of a vehicle. Especially telecommunications service providers such as Vodafone [47], Telekom [43] or Telefonica [42] offer OBD-II connectors in combination with a mobile data connection. Note, this also broadens the attack surface as it could provide remote access to the CAN (Controller Area Network) bus, transforming a perceived internal attack surface (needing physical access to the OBD-II port) into an external threat [17,49]. But apart from that, these service providers gain insights into an enormous amount of highly private data about the drivers' everyday life [24]. Among other things, this includes the driving style, routes and accurate driving times. Manufacturers and insurance companies as well as service providers from the business sector are already using the information provided as a basis for monetarizing the individual journeys of users. An example would be the reduction of car insurance if you have a particularly restrained and safe driving style. This monetarization can also work in the opposite direction. This means that drivers who drive very aggressively, for example, are punished with higher rates.

In summary, OBD-II communication creates two new threat categories:

1. Most obvious is the problem of malicious inbound flow from an adversarial dongle to the car
2. less obvious –judging from the limited amount of research so far– is the problem of personal data leakage in outbound flow from the car to the dongle

As there is no encryption or authentication on the CAN bus by default [18], not only inbound threats are very real and far from difficult to implement [18], but also the ability to tamper with data on its way from the car to the dongle is not prohibited due to the –by default– missing data origin-authentication.

1.1 Goals and Contributions

All this raises the question of whether it is possible to develop an effective and modular firewall-like filtering approach for data flows in both directions via the given interface. We therefore propose a rule-based approach suitable for the OBD-II interface and all affected protocols to protect the driver and his car from possible malicious dongles by modifying or rejecting data flowing in both directions.

The idea of something like a firewall in a car was proposed by Rivzi et al. where they showed a distributed firewall approach and put an inbound filter on each ECU in a vehicle [35]. However, our approach is specifically designed to manipulate traffic, especially the content of traffic from the car to the dongle. To the best knowledge of the authors this has not been done before and our approach still allows full or reduced operation of the dongle, unlike current commercially available blockers, which only offer an all-blocked or all-allowed approach. Let us stress that being able to manipulate traffic on OBD-II is an attack vector on its own, highlighting an absence of crucial security mechanisms for this communication channel. We exemplify the attack by building a Man-in-the-OBD-II interface to highlight how the data flow accessible via the OBD-II interface can be modified or restricted. One of the hurdles to be considered is the provision of the firewall for the end user. The objective is to make it as easy as possible for users to enter the firewall without any major entry hurdles. Therefore, in Sect. 4.7 we introduce a configurable policy language in the well-known JSON format, which eliminates this problem.

Our modular design works on the standardised OBD-II stack and we have chosen to implement both inbound filtering and outbound filtering for Controller Area Network protocol messages. CAN 2.0 specification was published in 1991 by Robert Bosch GmbH [36] and standardized by ISO 11898-1 [22] first in 2003, and is one of the most common used protocols for vehicular information exchange through dongles today. However, since there are many different communication mechanisms and protocols in the OBD-II stack, our general approach abstracts from the protocol itself. For brevity we focus in this work on CAN message filtering and rewriting in both directions, which can be seen as one module in our general OBD-II man-in-the-middle firewall concept. This makes it possible to easily add other protocols to our basic system with a functioning implementation for CAN. It means that our protocol-agnostic approach also enables manufacturer-specific solutions. The only hurdle that remains for anyone who wants to use the firewall is to configure the correct rules using our rule language. For this, the user must be able to understand the relevant important data, e.g. the message format within CAN, and know exactly which commands are sent. In the case of CAN, however, there is already a lot of current research that helps to reverse engineer messages [20,27,31].

1.2 Outline

In this work we focus on two aspects:

I: Identify the threats a security layer in between the OBD-II dongle and the car would solve

II: Implement such a security layer as a Man-in-the-Middle for the OBD-II and show how it can fool dongles to protect the car driver's privacy

We present an overview of related work in Sect. 2. In Sect. 3 we facilitate the standard threat modeling tools DREAD and STRIDE to analyze systematically the possible threats that can occur and, in the best case, can all be avoided by implementing our approach. We highlight the most important threats by means of this procedure. In Sect. 4 we present the architecture of our solution that abstracts the dongle from the actual vehicle network and shields the car as best as possible. In Sect. 5 we discuss briefly the implementation of our approach for the case of CAN messages. We provide a rule-based policy language. By means of different options and types within the rules it is possible to set the firewall filters for inbound and outbound data flows. We evaluate the impact on the identified threats and on selected dongles in Sect. 6. Finally, we conclude in Sect. 7.

2 Related Work

Work has been done on the vulnerabilities added by the dongles itself: The idea of analysing dongles' behaviour was discussed by Wen et al. [50]; they provide a comprehensive vulnerability analysis of 77 OBD-II dongles. In the paper the authors propose an automated tool called DongleScope to perform an analysis and to test the dongles. In the paper published by Yadav et al. [52] the authors give an overview of various security vulnerabilities and points of entry for malicious entities in vehicular systems.

The trove of information that can be gained from a car is shown in several works [16,24,28,39]; they show how to monitor automobiles, predict the condition of the internal hardware, detect driving behaviors and discover different anomalies.

In the remainder of the related work section we grouped the works by their view point on the information flow.

2.1 General Vehicular Security Concepts Describe the Threat of Unwanted Information Flow

In the paper by Bernardini et al. [9], eight security requirements and five safety requirements for vehicle systems are defined and explained. They also describe how existing systems and solutions such as the AUTOSAR architecture, LIN, FlexRay, MOST and Ethernet/BroadR-Reach are aligned and can be used to fulfil these requirements. Furthermore, the authors explain in detail which safety concepts are to be pursued in vehicle systems and which possible problems or limitations may arise.

Hoppe et al. [19] show that it is necessary to examine and possibly modify already existing security vehicle systems. Among other things, the authors show that the intrusion detection system does not fully protect cars from intruders. Even though this system is one of the newer ones in vehicle safety, according to Hoppe et al. some improvements need to be made to minimise the risk of attacks. This shows us that even current security concepts are often not fully developed and cannot offer complete protection. Therefore, our approach is to enable a security layer as simply as possible according to the plug and play concept.

Studnia et al. [41] have looked at fundamental problems related to car security. Among other things, they found that the computing power of a car is very limited and this can lead to problems when using strong cryptography within certain protocols. In addition, they figured out that car manufacturers must validate the software running on an ECU embedded within a vehicle and test it on a periodic basis to guarantee its integrity. An entire vehicle can become vulnerable if bugs remain in the vehicle system. These effects are of course reflected in the severity of the respective bug. In the event that a security flaw is utilised, it can require anywhere from several months to years for a patch to be installed for all of the specific cars that were already on the road. This implies that it is an extremely important task to prevent malicious code from entering the vehicle in the first place. Therefore, our solution is to safeguard the OBD-II interface and thus exclude possible attackers.

2.2 Filtering of Inbound Traffic Towards the Car's ECUs Exists

Wolf et al. [51] examined the prevailing architecture as well as the threats that are prevalent in contemporary vehicles. They discovered that the gateways built into the automotive network require the use of powerful firewalls. In addition to this, they stated that the firewall implemented in the gateways also need to possess rules that control access on the basis of the security relevance of the particular network.

While filtering approaches or firewall concepts for networks inside the vehicle do exist and are nothing completely unknown, the range of available research is very limited, especially compared to works on inter-vehicle networks like VANETS. Even less information is available about existing solutions for cars being deployed. For example NXP describes the need to protect the car's networked devices from unwanted outside traffic by a gateway for "filtering inbound and outbound network traffic based on rules, disallowing data transfers from unauthorized sources." [1]. NXP further states that a more fine-granular approach "[...] may include context-aware filtering" [1]. But often the exact mechanisms used as well as the security functions in real vehicles are not publicly published. Another manufacturer's solution is the "Central Gateway" for central in-vehicle communication from Bosch, which lists a firewall and an intrusion detection system on its product page [11]. However, neither the info PDFs nor the actual page list more precise details. Even when specifically asked at the responsible department, we were unable to get any further information about the security features mentioned. The company Karamba Security [23] in 2016 released a security architecture that acts as a gateway between a car's access points and critical networks/modules. Karamba calls it ECU Endpoint Security: Dropper Detection and Malware Prevention. To define factory policies, the developers had the idea of having a system embedded directly in the firmware. This is to prevent malicious code from infiltrating the system. Each ECU specifies its own policy and generates a so-called whitelist of permitted programm binaries, processes, scripts and network behaviour.

In academic literature Rizvi et al. presents this as a distributed approach for a firewall system in automobile networks [35]: Their system is focused to let only authorised packets reach an internal device using a Hybrid Security System (HSS) that uses many individual firewalls located in front of each module and at each electronic unit.

2.3 Commercial Available Approaches Towards Filtering and Securing the OBD-II Interface

Practical OBD devices which do offer such inbound filtering or just blocking all access are available on the market. The most critical feature touted as blockable by most entry filters is the use of "key duplicators and an accessible OBD-II socket" that would allow car thieves to generate new access codes thus obsoleting the original keys. This creates a safety barrier between the external devices and the data bus protecting vehicle functions against unauthorized access and manipulation. Using key duplicators and an accessible OBD-II socket, a car thief can easily generate new access codes, outflanking in a few moves the existing car alarm system. As can be clearly seen in Table 1, almost all approaches get delivered with only two modes supplied: always-deny (Off) or always-allow (On). Their focus is on preventing malicious senders' packets from reaching important devices in the vehicle by turning the CAN bus access off. Of course that would also completely block data that might travel in the other direction, but this privacy impact is not advertised. Furthermore, with the existing approaches, the use of an OBD-II dongle is not possible when activated, as absolutely no data is available for processing. Our approach closes this gap.

Table 1. Commercially available products' capabilities compared to our Man-in-the-OBD-II approach

Product	Operation types	Modes	Filtering	Method
Diagnostic BOX - OBD Blocker [44]	None	On/Off	None	MiM
Ampire CAN-BUS Firewall [2]	None	On/Off	None	MiM
Ampire OBD-Firewall [3]	None	On/Off	None	MiM
Paser Firewall OBD2 [30]	None	On/Off	None	MiM
Electronic anti theft OBD plug [46]	None	On/Off	None	MiM
AutoCYB [7]	None	Mounted/ Unmounted	None	Lock
CAN Hacker Diagnostic Firewall [14]	Unknown	Unknown	SIDs/PIDs	Unknown
Man-in-the-OBD-II	*reject, limit,* **replace**	*delay, pub_once,* **id_range, val_range**	*Individual*	*MiM*

3 Threat Modelling for OBD-II

We conducted a threat analysis for a commercial passenger car with an OBD-II dongle. Threats were categorized according to the STRIDE method [33]. We then use the DREAD method [38] to rate the seriousness of those threats.

3.1 Threats Following STRIDE

For brevity we limit this to the most relevant eight threats, which are used to draw a conclusion about the effectiveness or usability of our approach in Sect. 6.4.

T_α Malicious Device Plugs Directly into OBD-II
This threat is virtually impossible to prevent. As soon as the attacker has physical access to the interface, he can, for example, bypass all upstream hardware security modules (such as our firewall approach) or simply plug them off. T_α could only be effectively prevented by additional physical security mechanisms. One possibility would be to physically separate the hardware security modules from the accessible OBD-II interface in such a way that it is no longer possible for an attacker to either:

- Separate the module from the vehicle, or
- the use of the OBD-II interface becomes unusable as soon as the module is disconnected.

However, this would require some modification of the OBD-II standard and is not really feasible.

T_β Attacker Compromises a Running Service on Dongle
The threat class T_β is a class which includes threats that may or may not occur frequently, depending on the security standards applied in the development of the software running inside the dongle. The attack surface is very large due to the use of cloud services, an internet interface (e.g. by means of a SIM module in the dongle) or a Wi-Fi interface. The very fact that the service communicates with a cloud service via an API means that in addition to the service inside the dongle also the cloud services' endpoints must be regularly maintained and also updated. If this is not done, the risk of a possible attack increases further. In the following, there are two subcategories in the T_β class, namely:

I: Dongle refers to hardware where our firewall runs on
II: Dongle refers to the third-party hardware that plugs into our firewall

T_γ Attacker Can Send Arbitrary CAN Commands
This threat is enabled by either T_α or T_β. The attacker is thus able to control, for example, displays in the instrument cluster or the opening and closing of the electric windows. Generally speaking, it is possible for him to contact all ECUs that communicate via CAN and can be reached via the OBD-II interface. This enables a wide range of attack vectors.

T_δ Attacker Can Read Communication on CAN-Bus
No direct physical damage can be caused by this threat, as it is only a matter of reading access to the CAN bus. Nevertheless, the privacy of the respective user can be violated. It is possible to listen to all communication on the CAN bus that is accessible via OBD-II. By analysing the traffic, a variety of conclusions can be drawn. For example, statements can be made about the individual driving behaviour of each driver in road traffic.

T_ϵ Damaging ECUs by Executing Specific Commands
In order to damage individual ECUs by means of specific messages, the attacker needs a profound understanding of the respective vehicle system as well as the ECU to be damaged. In most cases, such an attack is not possible without extensive prior testing and analysis of the hardware. Of course, one can actively try to damage the hardware by sending harmful and unwanted CAN messages. However, a try-and-error approach offers little chance of success.

T_ζ Person Endangerment by Deactivating Safety Functions
This threat is a very dangerous one, as it may actively cause physical harm to people. On the one hand, an attacker could succeed in deactivating critical safety systems such as the airbag, the ABS (anti-lock braking system) or the ESP (electronic stability program). Thus, in the event of a driving manoeuvre where the respective system is required, a possible accident would be the result. Furthermore, an attacker could, for example, display the wrong speed. As an example, he could simply display a much lower speed than is actually being driven at the moment. Thus, the driver could be harmed if he is at that time in a certain road passage where maintaining a certain minimum speed is critical to safety.

T_η Permanent Infiltration of the Vehicle System by Uploading Malware
One goal during an attack or after a successful intrusion into a system is to make the access persistent. T_η describes the maintenance of unrestricted access even after possible malicious OBD-II devices have been removed. This requires in-depth knowledge of the ECUs available in the target vehicle. Possible targets are especially ECUs that have a memory module or dedicated firmware that can be overwritten.

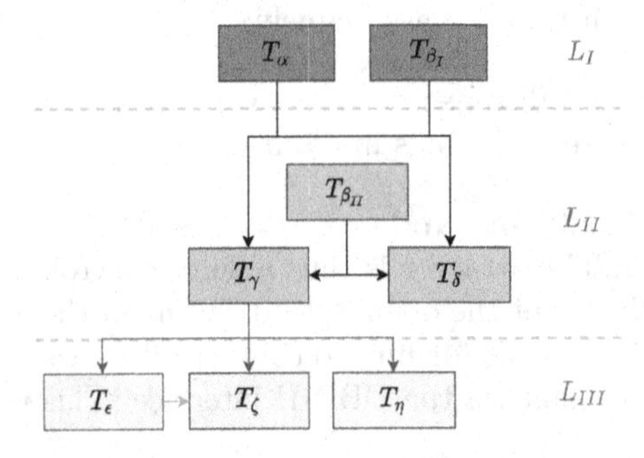

Fig. 1. Attack tree of specified threats

Table 2. DREAD rating of selected threats

	T_α	T_{β_I}	$T_{\beta_{II}}$	T_γ	T_δ	T_ϵ	T_ζ	T_η
D	3	3	3	3	2	3	3	3
R	3	2	2	2	2	2	2	1
E	3	2	2	2	2	1	2	1
A	3	3	3	3	3	3	3	3
D	3	2	2	3	3	2	3	2
DREAD risk	3.0	2.4	2.4	2.6	2.4	2.2	2.6	2.0

3.2 Results of OBD-II Threat Modeling

Based on the seven overall threat classes defined and explained above, we used the DREAD rating model [38] to roughly classify how big or serious the individual threats are. The individual values for each threat can be seen in Table 2. Levels range from low (1) to high (3) risk. When comparing the individual result values, it becomes clear that T_α is the highest rated threat with a risk rating of 3.0. The main reason for this is that T_α is a physical threat. With physical threats, the possibilities of an attacker are always greater than, for example, in the case of a remote only attack. The lowest rated threat is T_η with a risk rating of 2.0. This is because this type of attack requires a tremendous amount of knowledge and skill to execute. Usually a lot of research and testing needs to be done on the real physical devices/ECUs to even find vulnerabilities that make it possible to carry out the attack. It is also interesting that, with the exception of threat T_δ, the damage for all other threats is always rated at the maximum of three (*high*). The reason for this lies in the effect of the respective threats. T_δ describes the possibility to read messages of the CAN bus. This means that a possible attack can also be described as passive, since it never actively changes anything on the BUS or writes anything to it. Nevertheless, sensitive data (e.g. regarding the driver's privacy) can be collected by reading and possibly decoding some messages. Therefore, T_δ still has a rating of two, which means medium in terms of the DREAD rating system.

Since some threats enable other threats, we have created an attack tree of the seven selected threat classes. An analysis via attack trees provides a graphical, easy-to-understand modelling of threats. It helps us to classify the different attack possibilities of the OBD-II interface more precisely and to develop possible countermeasures to prevent such attacks. In Fig. 1 you can see this tree, which can also be divided into three hierarchies named L_I, L_{II} and L_{III}.

The first level is the basic threat hierarchy. This means that threats from higher levels are always based on threats from the basic level. The threats T_α and T_{β_I} are therefore needed to realise attacks with the threats from the layers below. Therefore, we also define these two threats as entry threats. In the best case, all entry threats can be eliminated or prevented or mitigated so that all further threats become obsolete or not quite as severe. The threats in our last level L_{III} are the most specific in terms of the knowledge required or the techniques used.

4 Architecture of the Man-in-the-OBD

In this section we describe the architecture and the individual abstract components, which we will then implement in Sect. 5. For brevity we only briefly explain the basic ideas of the respective components theoretically and show why we choose exactly this approach.

4.1 Producer/Consumer Scheme

The producer-consumer problem (also known as the bounded buffer problem) is a classic example of a multiprocess synchronisation issue, the first version of which dates back to Edsger W. Dijkstra in 1965. Nevertheless, there are now promising approaches in software development to efficiently eliminate this problem [26]. We have decided on such an approach (more details can be found in Sect. 5.1). A filtering approach is best realised with a buffer in which incoming messages are accumulated. Afterwards, they can be processed one after the other, depending on the queue. This model is ideal if you want to be as unrestricted as possible in the processing phase. Depending on the respective computing power, several producers or consumers can be started. In this way, load peaks can be easily absorbed. The modular approach can also be applied by means of differently implemented producers.

4.2 Modular Approach for Protocol Bindings

Since the producer/consumer scheme allows us to easily create several differently implemented producers, a uniform interface must be defined. This interface ensures that the responsible consumer can correctly process and forward the incoming messages. By means of this approach it is possible to support incoming messages of all protocols. With this method, a high-performance and efficient filtering is possible.

4.3 CAN-Bus Binding

In order to support the CAN protocol for our implementation, a connector is needed to receive messages as well as to be able to send filtered or processed messages again. For this purpose, already widely used libraries (e.g. [15,37]) as well as the common syntax for coding and decoding are used. Furthermore, it is desirable if the binding understands the so-called DBC format. DBC stands for CAN Database and is a proprietary format that describes the data structure over a CAN bus. A CAN DBC file is a text file that provides all the necessary information for decoding CAN bus raw data into physical values. If a DBC file is available for the respective manufacturer, the user can easily decode the CAN data streams and thus analyse them in order to manipulate or block them as desired in our firewall approach. This kind of provision can ensure a possibility to support as many manufacturers as possible. Analogous to this binding, a binding for ISO 9141 or SAE J1850 could also be written and integrated into our pipeline as a producer in order to support these protocols.

4.4 Processing Pipeline

A concurrent and multi-level data input and data processing pipeline is to be realised for the processing pipeline. It must also be possible to efficiently consume different sources, the so-called producers. It would also be desirable if there was the possibility to configure the processing pipeline with regard to the resources to be used. More precisely, the number of processes for producer and consumer as well as the concurrency and the batch size to be used. In Fig. 2 you can see an schematic overview of the pipeline. The blue circles in the figure symbolise the producers. As already mentioned, it will be possible to develop a producer for each protocol. So in the future there may be a producer for CAN, one for ISO 9141 and so on. There is a uniform interface to adhere to. The producers then send their messages to the concurrent and multi-stage data ingestion service. There, the incoming messages should be analysed and then asynchronously serialised and inspected. Here, serialisation refers to the application of the active rules and not to the quantity of messages.

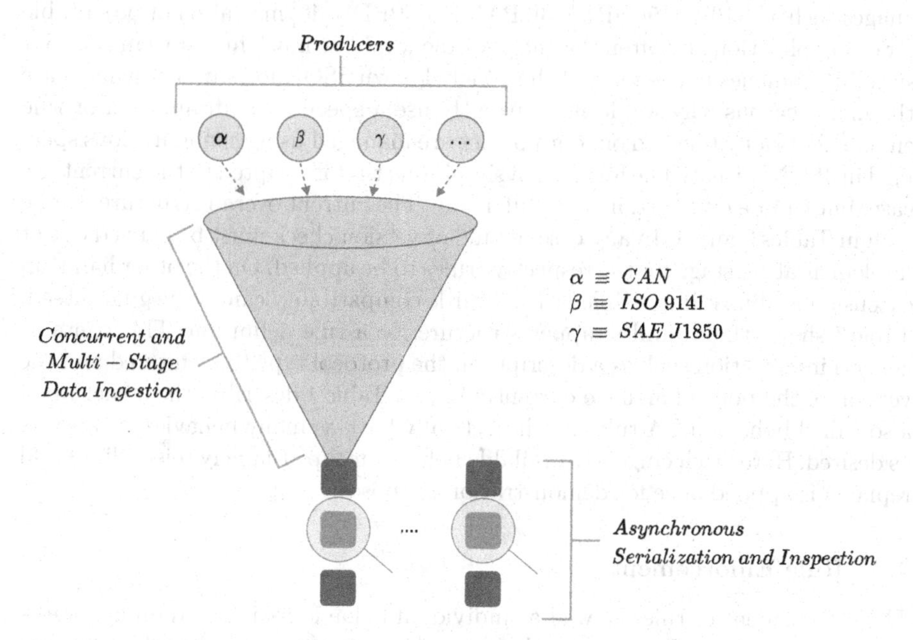

Fig. 2. Simplified representation of the processing pipeline

4.5 Serialization

After the incoming messages have been serialised and bundled into batches, the messages are to be checked for the active rules as efficiently as possible. Since the behaviours of a rule can be sorted according to their strictness in the restriction, the strictest behaviours should currently apply. This allows the individual behaviours to carry out the checks in parallel.

4.6 Data Storage

It should be possible to record the data as well. In the best case, data should be stored in a database in a uniform, reusable format. This ensures that the logged CAN messages can be easily searched or filtered for various purposes. Since the amount of CAN messages can be immense, there should be an option to deactivate or activate the permanent logging.

4.7 Policy Management

Policy management systems can be implemented either as specialised hardware or as software on general-purpose operating systems. However, the underlying idea is always the same. There is a set of defined rules that determine which packets the separated network can receive and how those packets are modified if necessary [10]. The best known firewall tools under Linux and Unix are *iptables* and *ipfw*. However, since these are far too extensive and complex for our current needs, we have opted for a simple implementation of our own. Other well-known policy languages such as RPSL [25], SRL [13], PAX [29], PFDL [40] may also not be suitable for our application and often the entry hurdle would be much higher than with our simplified policies in the form of the widely known JSON format. These facts are the main reasons why we do not currently use a specific rule framework or rule engine. As briefly touched on, our policies are managed using configurations specified in JSON format. The format is a simple one that is adapted to the current use case, but can be extended in a modular way. The current overall structure can be seen in Tables 3 and 4. In any case, a kind of version check must be carried out at implementation stage for the respective rules to be applied. On the other hand, an extension is almost impossible or backwards compatibility cannot be guaranteed. Table 3 shows the overall wrapper structure for a rule definition. This contains general information such as a description, the protocol type to be filtered and the version of the policy language currently in use. Table 4 describes the structure of a so-called behaviour. A rule can theoretically have as many behaviours capsules as desired. Here, each currently available behaviour type (namely reject, limit and replace) is applied once for demonstration purposes.

4.8 Rule Enforcement

The enforcement of rules as well as individual behaviours is based on an assessment of importance. This means that more important behaviours and rules outweigh less important ones. For this purpose, there is a special type rating, which is specified using the type property of the behaviours. As in Sect. 4.5, this has the great advantage that the individual behaviours can be executed in parallel after the initial filtering of the strictest rating and thus best fit our scalable approach.

Table 3. List of basic properties with their associated functionality

Property	Type	Description
name	<*String*>	Is just a simple naming of the individual rules for better distinction. The name does not have to be unique.
description	<*String*>	Briefly describes the created rule in a few words.
version	<*String*>	The version number specifies the version of the properties to be used.
protocol	<*Protocol-Type*>	Declares the protocol type to be used for the respective rule. Currently there is only $< CAN >$ as a declarable type.
behaviours	[<*Behaviour*>]	The behaviour field defines a list of all actions to be performed later during the execution of each rule.

Table 4. List of properties for a single behaviour inside a policy

Property	Type	Description
type	<*String*>	Currently, three different behaviour types are supported: – *reject* - Ignores all messages with the defined identifier and associated value – *limit* - Limits all accruing values of a message from the defined identifier by means of a predefined value – *replace* - Always exchanges all message values of a given identifier with the given value
identifier	<*String*>	Defines the identifier of the CAN message present on the bus
value	<*String*>	Determines the data payload to be used for the respective set type
The following properties are optional and do not have to be set		
delay	<*Integer*>	If the delay property is set, all messages that fall below the specified behaviour will be delayed. The value is given as an integer value and defines the delay time in milliseconds.
pub_once	<*Boolean*>	Allows messages in the scope of the behaviour to be allowed only once per system start. Once the message has been read once, it is whitelisted and then not forwarded. By default, the value is set to *false*.
id_range	<*String*>	By means of the identifier range, the behaviour value range to be enforced can be extended.
val_range	<*String*>	Allows messages in the scope of the behaviour to be allowed only once per system start. Once the message has been read once, it is whitelisted and then not forwarded. By default, the value is set to *false*.

5 Implementation

Next we describe all the components that run our developed approach in the background. This includes the processing of incoming messages, the used producer & consumer approach, the storage of individual messages as well as our filter module.

5.1 Producer/Consumer Solution

As already described in Sect. 4.1, our approach should benefit from the so-called producer consumer construct and thus make a modular approach more feasible. This is one of the disciplines in which Elixir can demonstrate all its abilities and advantages in the best possible way. To build our solution, we use the library called *Broadway* [12].

The library can be used to create concurrent, multi-stage data input and data processing pipelines. It is also possible to implement your own producers, which is perfect for our use case. Depending on the use case, it can make a lot of sense to summarise the processed messages as a so-called batch before the actual publication. While we don't need the batchers to communicate with an extraneous API, it allows us to process the storing of CAN messages in an encapsulated way and thus have no runtime loss for publishing already filtered CAN messages. This allows for increased throughput and consequently improved overall performance of our pipeline. Batches are simply defined via the configuration option. The configuration is of course adaptive and can be extended very easily if required. A schematic representation of how our current testing pipeline looks is shown in Fig. 3.

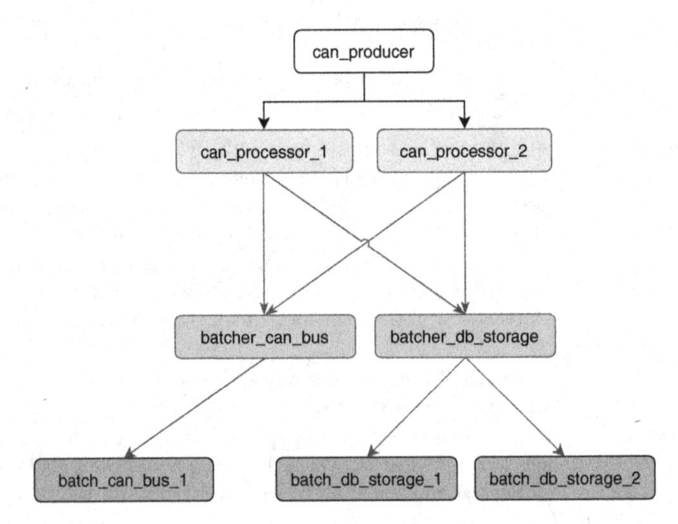

Fig. 3. Representation of producer, processor and batcher pipeline

5.2 Storing of CAN-Messages

The storage of CAN messages can be set via the web user interface. By default, CAN messages are not saved. However, if the option is activated, all filtered messages are stored in a PostgreSQL database in the `batcher_db_storage`. Since the storage is carried out in one of the batcher processes, all storage operations can be carried out completely independently of the runtime of the actual filter pipeline. PostgreSQL was chosen because the *Phoenix framework* uses it as the default database. The advantage of this, however, is that the adapter responsible for the connection to the database can simply be exchanged. This makes it easy to switch to a Time Series Database (TSDB) such as InfluxDB [21] or Riak [34]. TSDBs are suitable because CAN messages are time-distributed data. Since the storage and further processing of past CAN bus data is not relevant for our work, the use of PostgreSQL is sufficient for our purposes.

In addition to the raw data of the identifier and the payload which are stored using the `byte` data type, we also store the encoded data as `varchar(255)`. This has the advantage that we can display the data in a simple and user-friendly way in the front-end without having to decode a large number of data sets each time.

5.3 Pipeline Benchmarks

In Table 5 you can see the measurement results for the runs of our pipeline with different numbers of behaviours. Our firewall runs on a Raspberry Pi 4 with 8 GB RAM. The measurements generally show that our filtering pipeline only requires a small amount of computing time and therefore has no impact on the usability of the dongles. The average runtime of the pipeline increases with the increase in the number of behaviours defined in a rule. In general, the additional execution time has no effect on the correct functioning either way, as the dongles are inherently time uncritical (more on this in Sect. 7).

Table 5. Results of the run time comparison of the single measurements

Measurement	Iterations per second	Average	Deviation	Median	Minimum	Maximum	Sample Size
3 behaviours	243.87	4.10 ms	±9.71%	4.08 ms	3.05 ms	6.79 ms	1218
30 behaviours	225.89	4.43 ms	±11.89%	4.36 ms	3.29 ms	10.23 ms	1218
300 behaviours	164.58	6.08 ms	±11.90%	6.03 ms	4.45 ms	11.07 ms	823
3000 behaviours	83.22	12.02 ms	±7.44%	12.02 ms	9.55 ms	14.68 ms	416
30000 behaviours	16.74	59.72 ms	±7.09%	59.10 ms	51.62 ms	69.53 ms	84

6 Evaluation of Impact on Threats and Existing Dongles

To evaluate we have tested our approach with real software currently available on the market; we report the tests of the RYD-Box [45] and the Volkswagen (VW) Data Plug [48] in this paper. We ran two different test cases both were carried out with a Volkswagen Golf model 7: In case $T1$ messages are blocked (Table 6 contains detailed information about the blocked messages). In case $T2$, we manipulate messages' contents (for details see Table 7). In addition to the respective PID (process identifier) of the message, we have also included the minimum and maximum values defined in the standard and the firewall response defined by our rule in the tables, e.g. `blocked` in case $T1$ and limiting the speed value to a maximum of `100` in $T2$. Finally, we also theoretically discuss which threats identified using STRIDE and sorted using DREAD can be mitigated by our approach.

6.1 Testing the RYD-Box

First we ran $T1$ and blocked CAN requests for both speed and current mileage using our firewall. Then we did a test drive and checked the data displayed to us in the app. Here we were still shown the last known mileage, which no longer matched after the drive with the firewall activated. We had tested the dongle beforehand without the firewall, so the service already had access to our mileage at that time. This shows that the mileage is stored in the cloud at certain times of the respective car, but also that it is possible to successfully block the transmission of the mileage with our approach and still use the remaining functions of the dongle. We then displayed the speed history of our recorded journey. There we noticed that despite blocking the messages for the current speed on the OBD-II interface, the approximate driving speed was also tracked in the app. Our first assumption was that the app calculated the speed using the GPS data of the smartphone. However, after another drive without a smartphone in the car, we found out that the RYD-Box has both a GSM module with a separate SIM card and a GPS sensor. This enables the service to approximate the speed using GPS. Alternatively, additional sensors such as an acceleration sensor or a gyroscope could be installed to collect driving statistics. However, we have not yet disassembled the dongle and can therefore only list additional possibilities to the GSM and GPS modules.

Afterwards, we reconfigured our firewall for $T2$ so that it always returns 200000 km when the current mileage is requested. In addition, we have limited the current speed to 50 km/h in a rule using behaviour. Then we did the same test drive as for $T1$ and compared the results in the app again. As we expected, the speed limit had no effect on our tracked distance. We simply assume that the application does not use the current speed transmitted via CAN for the calculations or displays within the app for a journey. However, we could see modified mileage, namely the value of 200000 km as defined in the behaviour of our rule, becoming displayed in RYD's smartphone app, without any hint on it being modified.

6.2 Testing the VW Data Plug

We blocked the CAN requests to the OBD-II gateway for $T1$ using our firewall for the current mileage and the current speed. After connecting our mobile phone via bluetooth to the dongle and the app, we were able to display data from our vehicle. However, the mileage could not be displayed because it was successfully blocked by our firewall. Once again, we drove a test lap for the recording of a journey. Again, just like with the RYD-Box, a speed could be read out in the app afterwards with the Volkswagen Data Plug, despite the blocking by our firewall. However, here again we have the same case as in the test with the RYD-Box. The speed for the saved journeys is simply not displayed over the speed that can be requested via OBD-II. In contrast to the RYD-Box, the VW Data Plug from Volkswagen does not have a GSM or GPS module, but the application must have access to the current location and the GPS services of the smartphone in order to function at all. For this reason, no test was possible without a smartphone in the vehicle. Due to the compact and smaller design of the VW Data Plug, we can assume that no additional sensors are installed there, and only the sensors installed in the smartphone (such as the acceleration sensor or the gyroscope) are used to evaluate the driving behaviour.

For the $T2$, our firewall was again configured in the same way as it was above during the test for the RYD-Box. After our new round of tests, we were able to prove exactly the same behaviour as the RYD-Box showed. We were able to see our manipulated vehicle mileage within the We Connect application, but the manipulation of the speed had no effect, as the speed is approximated by GPS as explained before.

Table 6. Overview of CAN-Messages to be blocked in $T1$

PID (hex)	min_value	max_value	Firewall response	Description
0C	0	255	*<blocked>*	Vehicle speed
A6	0	429,496,729.5	*<blocked>*	Odometer

Table 7. Overview of CAN-Messages to be manipulated in $T2$

PID (hex)	min_value	max_value	Firewall response	Description
0C	0	255	*<min_value>* to 100	Vehicle speed
A6	0	429,496,729.5	200,000.0	Odometer

6.3 Evaluation of Existing Dongles

These two examples in Sect. 6.1 and 6.2 are only intended to show the feasibility and do by no means represent a full analysis of all car manufacturers in combination with different dongles. Because our approach is protocol agnostic and also not dependent on the underlying message format, it can be assumed that our OBD-II firewall will work with all possible combinations. Our main takeaway from testing against real existing dongles is that our approach is surprisingly easy to apply in practice. This means that no message origin authentication features (like digital signatures on message contents [8,32]) on the CAN bus are used, as the dongle does not detect the manipulation of message's content in test case $T2$.

Additionally we gained interesting deeper insights into the functionality of the two dongles. We contemplate a closer look at the dongles in combination with the related smartphone application, such as in the case of the Volkswagen Data Plug, because we found that this combined setup additionally used information collected from the smartphone's sensors; these might also be tampered with - albeit in different ways: We consider it quite conceivable to trick the application corresponding to the dongle to the extent of feeding it false journeys with faked speeds. To do this, you would have to isolate the application to run it within an emulator where you can influence the position information and the gyroscope's readings.

6.4 Evaluation of Threat Mitigation

Finally we check whether our approach prevents or at least mitigates the threats T_α to T_η as identified in Sect. 3. With regard to the threats T_α and T_{β_I}, our approach unfortunately cannot do anything. In general, it is almost impossible to achieve hundred percent security against physical attacks or manipulations. With regard to the β threat class, however, you can try to make your services within the vehicle as secure as possible and close any security gaps discovered with regular updates. However, if the attacker has physical access, often even the best systems cannot be protected. Direct and unlimited access to low-level debug interfaces such as those provided by JTAG (Joint Test Action Group) and SWD (Serial Wire Debug) would make it possible to take complete control of the device, e.g. stop and change code execution, access memory and registers, or even dump the firmware.

However, T_γ is preventable by means of our approach. With the help of the rules and the behaviours, it can be defined within the firewall which messages are allowed through and which are not. This means that it is no longer possible for an attacker to simply send arbitrary messages.

Likewise, the threat T_δ can be prevented by configuring the firewall for the desired permitted communication using the method described above. Our approach not only allows or blocks certain messages, but also enables to modify the data payload for defined messages before publishing them to the OBD-II device.

Since the threats T_γ and T_δ can be prevented by our approach, the threats based on T_γ can also be avoided automatically. Thus, threats T_ϵ, T_ζ and T_η are also eliminated. This means that our firewall approach successfully prevents all threats on the hierarchies L_{II} and L_{III} (see Fig. 1).

7 Conclusion

While it is yet not really widespread to install filtering approaches such as firewalls in vehicle systems at the OBD-II outgoing interface, this can become a crucial interconnection point between the car and third party dongles in the future. The need for such solutions will grow in the coming years, as more and more different approaches and products will come onto the market to make existing cars even smarter. And the OBD-II interface is clearly the standard entry point for realising such smart car approaches. Manufacturers agree that security must be added to CAN or OBD-II, but dongles available on the market today can be easily fooled by manipulated messages. Following the threats we have systematically defined in Sect. 3, the proposed Man-in-the-OBD shows that without any means to authenticate messages from the car's electronic control units, e.g. the car's current speed, dongles connected to the OBD-II interface can be fooled. We are the first to show that a Man-in-the-OBD-II interface is able to serve the dongle manipulated values without it noticing the falsification. We exploit the missing authentication to protect the car's drivers privacy. This is of course a benign application, used maliciously the falsified information provided to the environment of the car could have negative consequences. During the test cases, we were also able to test whether our firewall without rules has an effect on the functionality of the dongles used. This was not the case which can be easily explained by the fact that our approach in normal cases without active roles is simply two physically separated CAN-BUS systems forwarding all messages between the buses.

L_I-threats are directly attacking the dongle, like T_α and T_{β_I}. Those can of course not be mitigated by our firewall, because our firewall simply is a dongle. However, we show that threats of hierarchies L_{II} and L_{III} can be effectively prevented by policing the OBD-II interface. This is already an important step to protect against external malicious devices. We have implemented it and prototypically showed that it works for the CAN protocol. However, our architecture is general, and our approach could be used even for other interfaces; our modular approach allows to extend it to other protocols.

For an increased privacy we showed that the connected dongles providing data for cloud services to monitor the car via a mobile phone application are not affected by the small delay caused by our filtering or manipulation of the data traffic.

7.1 Future Work

Due to the diversity of protocols available via OBD-II, we limited ourselves to the CAN protocol. It would definitely be an enrichment for our firewall if the remaining protocols accessible via OBD-II were also covered by our firewall.

Furthermore, the OBD-II interface is the most exposed interface, but it would be valuable to research a firewall or man-in-the-middle attack at other interfaces; for example the infotainment system. Technologies such as AndroidAuto [4] and Apple CarPlay [6] are already finding their way into vehicles. But also completely independent Android systems are delivered with the so-called Android Automotive OS [5]. If apps would have vulnerabilities or contain malicious code it would be fatal if such apps could suddenly gain control over the vehicle. Therefore, an adapted firewall approach similar to our current one could also be very useful to isolate the car from the infotainment system. Another point that should be examined in future work is how to implement our developed approach on less expensive devices than the Raspberry Pi. If applicable, approaches should also be considered here that shift the actual computations to a central processing unit provided by the car, as an example.

References

1. Gateway, A.: A Key Component to Securing the Connected Car. Whitepaper, NXP Semiconductors (2018)
2. Ampire: CAN-Firewall (2022). https://www.ampire.de/-WFS300-BT.htm?SessionId=&a=article&ProdNr=%5FWFS300-BT&p=1857
3. Ampire: OBD-Firewall (2022). https://www.ampire.de/Product-Archive/Ampire/Theft-protection/AMPIRE-OBD-Firewall-without-harness-.htm?shop=ampire_en&SessionId=&a=article&ProdNr=%5FOBD-FIREWALL&p=1857
4. Android: Android Auto (2022). https://www.android.com/auto/
5. Android: Android OS (2022). https://source.android.com/devices/automotive
6. Apple: Apple Carplay (2022). https://www.apple.com/ios/carplay/
7. AutoCYB: Vehicle cybersecurity lock (2022). https://autocyb.com/product/autocyb-vehicle-cybersecurity-lock/

8. Bauer, J., Staudemeyer, R.C., Pöhls, H.C., Fragkiadakis, A.G.: ECDSA on things: IoT integrity protection in practise. In: Lam, K., Chi, C., Qing, S. (eds.) Information and Communications Security - 18th International Conference, ICICS 2016, Singapore, November 29–December 2, 2016, Proceedings. LNCS, vol. 9977, pp. 3–17. Springer, Cham (2016). https://doi.org/10.1007/978-3-319-50011-9_1
9. Bernardini, C., Asghar, M.R., Crispo, B.: Security and privacy in vehicular communications: challenges and opportunities. Vehicul. Commun. **10**, 13–28 (2017). https://doi.org/10.1016/j.vehcom.2017.10.002
10. Bodei, C., Degano, P., Galletta, L., Focardi, R., Tempesta, M., Veronese, L.: Language-independent synthesis of firewall policies. In: 2018 IEEE European Symposium on Security and Privacy (EuroS&P), pp. 92–106 (2018). https://doi.org/10.1109/EuroSP.2018.00015
11. Bosch: Bosch central gateway (2022). www.bosch-mobility-solutions.com/en/products-and-services/passenger-cars-and-light-commercial-vehicles/connectivity-solutions/central-gateway/
12. Broadway: Concurrent, multi-stage tool for building data ingestion and data processing pipelines (2022). https://hexdocs.pm/broadway/Broadway.html
13. Brownlee, N.: SRL: A Language for Describing Traffic Flows and Specifying Actions for Flow Groups. RFC 2723 (1999). https://doi.org/10.17487/RFC2723
14. CAN Hacker: Automotive diagnostic firewall (2022). https://canhacker.com/projects/obd2-diagnostic-firewall/
15. cantools: Can bus tools (2022). https://cantools.readthedocs.io/en/latest/
16. El Basiouni El Masri, A., Artail, H., Akkary, H.: Toward self-policing: detecting drunk driving behaviors through sampling can bus data. In: 2017 International Conference on Electrical and Computing Technologies and Applications (ICECTA), pp. 1–5 (2017)
17. Gmiden, M., Gmiden, M.H., Trabelsi, H.: An intrusion detection method for securing in-vehicle can bus. In: 2016 17th International Conference on Sciences and Techniques of Automatic Control and Computer Engineering (STA), pp. 176–180 (2016). https://doi.org/10.1109/STA.2016.7952095
18. Groza, B., Murvay, P.S.: Security solutions for the controller area network: bringing authentication to in-vehicle networks. vol. 13, pp. 40–47 (2018). https://doi.org/10.1109/MVT.2017.2736344
19. Hoppe, T., Kiltz, S., Dittmann, J.: Security threats to automotive can networks – practical examples and selected short-term countermeasures. In: Proceedings of the 27th International Conference on Computer Safety, Reliability, and Security, pp. 235–248. SAFECOMP 2008, Springer, Berlin (2008). https://doi.org/10.1007/978-3-540-87698-4_21
20. Huybrechts, T., Vanommeslaeghe, Y., Blontrock, D., Van Barel, G., Hellinckx, P.: Automatic reverse engineering of can bus data using machine learning techniques, pp. 751–761 (2018). https://doi.org/10.1007/978-3-319-69835-9_71
21. InfluxData: The time series data platform where developers build IoT, analytics, and cloud applications (2022). https://www.influxdata.com
22. Road vehicles - Controller area network (CAN) - Part 1: Data link layer and physical signalling. Standard ISO 11898–1, International Organization for Standardization, Geneva, CH (Dec 2015)
23. Karamba Security: Homepage. https://karambasecurity.com
24. Keegan, J., Ng, A.: Who is collecting data from your car?). https://themarkup.org/the-breakdown/2022/07/27/who-is-collecting-data-from-your-car

25. Kisteleki, R., Haberman, B.: Securing Routing Policy Specification Language (RPSL) Objects with Resource Public Key Infrastructure (RPKI) Signatures. RFC 7909 (2016). https://doi.org/10.17487/RFC7909, https://www.rfc-editor.org/info/rfc7909
26. Lamport, L.: A New Solution of Dijkstra's Concurrent Programming Problem, pp. 171–178. Association for Computing Machinery, New York, NY, USA (2019). https://doi.org/10.1145/3335772.3335782
27. Marchetti, M., Stabili, D.: Read: Reverse engineering of automotive data frames. IEEE Trans. Inform. Foren. Secur. 1–1 (2018). https://doi.org/10.1109/TIFS.2018.2870826
28. Nirmali, B., Wickramasinghe, S., Munasinghe, T., Amalraj, C.R.J., Bandara, H.M.N.D.: Vehicular data acquisition and analytics system for real-time driver behavior monitoring and anomaly detection. In: 2017 IEEE International Conference on Industrial and Information Systems (ICIIS), pp. 1–6 (2017)
29. Nossik, M., Richardson, M., Welfeld, F.J.: PAX PDL - a non-procedural packet description language. Internet-Draft draft-nossik-pax-pdl-00, Internet Engineering Task Force (1998). https://datatracker.ietf.org/doc/draft-nossik-pax-pdl/00/. (work in Progress)
30. Paser: Firewall OBD2 (2022). https://automotive.paser.it/en-gb/Paser/Firewall-OBD2-card-included-p1069m11.html
31. Pesé, M., Stacer, T., Campos, C., Newberry, E., Chen, D.: Librecan: automated can message translator, pp. 2283–2300 (2019). https://doi.org/10.1145/3319535.3363190
32. Pöhls, H.C., Petschkuhn, B.: Towards compactly encoded signed IoT messages. In: 22nd IEEE International Workshop on Computer Aided Modeling and Design of Communication Links and Networks, CAMAD 2017, 19–21 June 2017, pp. 1–6. IEEE, Lund, Sweden (2017). https://doi.org/10.1109/CAMAD.2017.8031622
33. Praerit Garg, L.K.: The stride threat model (2020). https://docs.microsoft.com/en-us/previous-versions/commerce-server/ee823878(v=cs.20)?redirectedfrom=MSDN
34. Riak: the world's most resilient nosql databases (2022). https://riak.com
35. Rizvi, S., Willett, J., Perino, D., Vasbinder, T., Marasco, S.: Protecting an automobile network using distributed firewall system. In: Proceedings of the Second International Conference on Internet of Things, Data and Cloud Computing. ICC 2017, Association for Computing Machinery, New York, NY, USA (2017). https://doi.org/10.1145/3018896.3056791
36. Robert Bosch GmbH: CAN Specification (1991). version 2.0
37. Sholik, A.: Porcelain (2022). https://github.com/alco/porcelain
38. Shostack, A.: Experiences threat modeling at microsoft (2008)
39. Srinivasan, A.: IoT cloud based real time automobile monitoring system. In: 2018 3rd IEEE International Conference on Intelligent Transportation Engineering, pp. 231–235 (2018)
40. Strassner, J., Schleimer, S.: Policy Framework Definition Language. Internet-Draft draft-ietf-policy-framework-pfdl-00, Internet Engineering Task Force (1998). https://datatracker.ietf.org/doc/draft-ietf-policy-framework-pfdl/00/. (work in Progress)

41. Studnia, I., Nicomette, V., Alata, E., Deswarte, Y., Kaâniche, M., Laarouchi, Y.: Security of embedded automotive networks: state of the art and a research proposal. In: ROY, M. (ed.) SAFECOMP 2013 - Workshop CARS (2nd Workshop on Critical Automotive applications: Robustness & Safety) of the 32nd International Conference on Computer Safety, Reliability and Security, p. NA. Toulouse, France (2013). https://hal.archives-ouvertes.fr/hal-00848234
42. Telefonica: o2 car connection von telefonica (2015). https://www.techstage.de/test/Im-Test-o2-Car-Connection-von-Telefonica-Germany-2572330.html
43. Telekom: Carconnect: Lösung für vernetztes fahren (2019). https://www.telekom.de/smarte-produkte/iot?wt_mc=alias_301_smarte-produkte/iot/carconnect
44. The Diagnostic Box: OBD blocker (2022). https://thediagnosticbox.com/product.php?pc=OBD+Blocker
45. ThinxNet GmbH: Ryd box (2022). https://de.ryd.one
46. UniversClub: Electronic anti-thefts systems (2022). https://www.universclub.com/en/buy/cat-electronic-anti-thefts-systems-2164.html
47. Vodafone: V-auto: So einfach wird dein auto zum connected car (2019). https://www.vodafone.de/featured/innovation-technologie/v-auto-so-einfach-wird-dein-auto-zum-connected/
48. Volkswagen: Volkswagen data plug from texa (2022). https://www.volkswagen.de/de/konnektivitaet-und-mobilitaetsdienste/konnektivitaet/we-connect-go.html
49. Wang, Q., Sawhney, S.: Vecure: a practical security framework to protect the can bus of vehicles. In: 2014 International Conference on the Internet of Things (IOT), pp. 13–18 (2014). https://doi.org/10.1109/IOT.2014.7030108
50. Wen, H., Chen, Q.A., Lin, Z.: Plug-n-pwned: comprehensive vulnerability analysis of OBD-II dongles as a new over-the-air attack surface in automotive IoT. In: 29th USENIX Security Symposium (USENIX Security 20), pp. 949–965. USENIX Association (2020). https://www.usenix.org/conference/usenixsecurity20/presentation/wen
51. Wolf, M., Weimerskirch, A., Paar, C.: Security in automotive bus systems (2004)
52. Yadav, A., Bose, G., Bhange, R., Kapoor, K., Iyenger, N.C.S.N., Caytiles, R.: Security, vulnerability and protection of vehicular on-board diagnostics 10. Int. J. Secur. Appl. 405–422 (2016). https://doi.org/10.14257/ijsia.2016.10.4.36

Mapping the Security Events to the MITRE ATT&CK Attack Patterns to Forecast Attack Propagation (Extended Abstract)

Roman Kryukov[1], Vladimir Zima[1], Elena Fedorchenko[2(✉)], Evgenia Novikova[2], and Igor Kotenko[2]

[1] A.F. Mozhaysky Military-Space Academy, Zhdanovskaya str. 13, St. Petersburg 197198, Russia
roman682@yandex.ru, zima_vka@mil.ru

[2] St. Petersburg Federal Research Center of the Russian Academy of Sciences (SPC RAS), St. Petersburg Institute for Informatics and Automation of the Russian Academy of Sciences, 14-th Liniya, 39, St. Petersburg 199178, Russia
{doynikova,novikova,ivkote}@comsec.spb.ru

Abstract. Modern information systems generate a lot of events. Analysis of the events allows detecting malicious activity within the system. There are a lot of event correlation techniques intended for the detection of cyber security incidents and different types of cyber attacks, as well as there are a lot of techniques for multi-step attack modeling. At the same time, most modern security event management solutions do not allow mapping the detected security incidents to the specific stage of the targeted multi-step cyber attack, forecasting the next steps of the cyber attack, and selecting the proactive responses automatically. In this paper the technique to map the detected incidents to the stages of the targeted cyber attacks is proposed. The technique is based on the set of correlation rules "Emerging Threats" for events correlation to get cyber security incidents and on the set of "Targeted Attack Analyzer (Indicators Of Attack)" rules describing security incidents (signatures) using Sigma language and integrated with the MITRE ATT&CK database. The developed technique allows mapping the events detected in the system under analysis to the MITRE ATT&CK attack patterns and in prospect forecasting the targeted cyber attack development and automatically responding against the detected cyber security incidents. The technique is implemented using Python language and tested to demonstrate mapping of the detected incidents to the known attack patterns using the deployed test environment.

Keywords: Security events · Cyber security incidents · Event correlation · Correlation rules · Signature · Cyber attack · Attack pattern · MITRE ATT&CK · Targeted attack analyzer · Indicators of attack · Emerging threats

© The Author(s), under exclusive license to Springer Nature Switzerland AG 2022
W. Li et al. (Eds.): ADIoT 2022, LNCS 13745, pp. 165–176, 2022.
https://doi.org/10.1007/978-3-031-21311-3_10

1 Introduction

In modern information systems, a large number of different processes including malicious processes take place. Security information and event management systems (SIEM) were introduced to understand what is happening in the system under analysis, as well as to detect, forecast and timely prevent the development of malicious activities. These systems allow the gathering of information on the operation of different systems (event logs). To detect malicious activity and cyber security incidents, the correlation of events from various sources is used.

The term 'correlation' originates from statistics – "When the relationship is of a quantitative nature, the appropriate statistical tool for discovering and measuring the relationship and expressing it in a brief formula is known as correlation." [7]. In information security, the event correlation means establishing a dependency between the events that relate to the same cyber security incident [11]. Key stages of the event correlation process are normalization, preprocessing, anonymization, aggregation, filtering, correlation itself, and prioritization [11]. The results of the event correlation including the detected cyber security incidents can be used to attribute the attacker [8], forecast the attack development, especially for targeted cyber attacks, and select appropriate countermeasures. We outline the targeted cyber attacks because of their features such as a specific target that can be achieved using different methods, multi-step execution, distribution in time, and complexity. The cyber attacks usually incorporate several standard stages, namely, reconnaissance, resource development, initial access, execution, persistence, privilege escalation, defense, evasion, credential access, and discovery [2]. Detection of the targeted cyber attack at an early stage and correct forecasting of the attack goal allow avoiding the attack's success and impact from its successful implementation. Thus, the task of the accurate mapping of the detected incident to the attack stage is highly relevant.

Currently, there are a lot of event correlation techniques for the SIEM systems. They use manual, supervised and unsupervised methods for the event correlation. Manual methods use the rules or attack signatures generated by the experts, supervised methods use the training dataset to construct dependencies between the events and their features, and unsupervised methods do not use prior knowledge [12]. There are also a lot of information security monitoring tools that implement such techniques. It should be noted that most security monitoring systems allow detecting the cyber security incidents using correlation techniques but do not allow mapping the detected security incidents to the targeted cyber attack stages, and, as a result, do not allow forecasting the next steps of the cyber attack and selecting the proactive responses automatically. To fill this gap, the authors propose the technique based on the set of "Emerging Threats" [3] correlation rules for events correlation to form cyber security incidents and on the set of "Targeted Attack Analyzer (Indicators Of Attack)" rules (TAA (IOA) rules) [1] to map the detected incidents to the stages of the targeted cyber attacks. The TAA (IOA) rules are the rules describing a suspicious behavior in the system (signature) that could indicate a targeted attack (security incident) [1] and can be checked in real-time. The authors propose using the open set of the TAA (IOA) rules specified using Sigma language [4] and integrated with the MITRE ATT&CK database. The IOA allows mapping the security incidents (signatures) to the attack stages specified in the MITRE ATT&CK database [2], namely, reconnaissance, resource development, initial access, execution, persistence, privilege escalation, defense, evasion, credential access, and discovery.

The proposed technique allows detecting the security incidents on the basis of the security events and mapping them to the IOA that, in their turn, are mapped to the MITRE ATT&CK tactics and techniques. In prospect, it will allow forecasting of the next attack steps and timely response against cyber attacks.

Contribution. The main contribution of the paper consists in the development of a comprehensive technique and tool that allow detecting the cyber security incidents on the basis of event logs and mapping the detected cyber security incidents to the stages of the cyber attacks according to the MITRE ATT&CK.

Novelty. The novelty of the proposed solution consists in the introduced technique as a whole and its specific elements, namely, the mathematical apparatus used to represent correlation rules and TAA (IOA) rules, and the algorithm used to map the cyber security incidents to the attack stages.

The paper is organized as follows. Section 2 analyzes the related research in the area. Section 3 introduces the proposed technique. Section 4 describes the implementation of the proposed technique and the conducted experiments. Section 5 contains the discussion, conclusion, and future work directions.

2 Related Research

Depending on the way of forming the knowledge on a security event it is possible to outline manual, supervised and unsupervised correlation methods. The manual methods use the rules or attack signatures generated by experts [12]. They include rule-based and pattern-based methods, codebook (case) based methods, model based reasoning, and graph based methods. The supervised and unsupervised methods use different machine learning methods and, therefore, are considered often the black box solution. The supervised methods use the training dataset to generate the rules for triggering alerts, while the unsupervised methods do not use any prior knowledge, and are based on statistical methods or deep learning techniques [12].

SIEM systems usually use several event correlation methods. For example, Splunk Enterprise Security[1] uses neural network based correlation that allows detecting anomalies in the event stream using the trained neural network.

[1] https://www.splunk.com/en_us/products/enterprise-security.html.

QRadar SIEM[2], HP ArcSight Security Intelligence[3], and MaxPatrol SIEM[4] use rule-based correlation methods.

While all these systems allow detecting security incidents and their types, only a few of them provide mapping to the indicators of attacks. For example, PT Network Attack Discovery component[5] from Positive Technologies allows automated mapping of the detected incidents to a set of the attack techniques and tactics. SIEM QRadar from IBM includes QRadar Use Case Manager that allows creating own rules for mapping detected incidents to specific tactics and techniques.

The most commonly used public knowledge-based repository of adversary tactics and techniques is the MITRE ATT&CK repository. As of June 2022, MITRE ATT&CK provides more than 620 attack techniques for enterprise information platforms. Apart from the repository of the real word adversarial techniques the MITRE ATT&CK project provides a mapping of the tactics, techniques, and procedures (TTPs) to attack stages that could be used to forecast attack deployment, reveal missed attack steps and select appropriate countermeasures.

For example, in [9], a framework for attack path analysis that utilizes the knowledge graph constructed on the basis of the MITRE ATT&CK matrix is proposed. The authors develop several algorithms for forecasting and discovering missing attack steps based on graph analysis. A similar problem is investigated in [13], Nisioti et al. suggest using MITRE ATT&CK knowledge base to support the investigation process step-by-step by producing possible recommendations based on the actions performed by the analyst and attributes and relations between attack tactics and techniques.

In [5] a proactive methodology for a system security assessment is presented, it includes an adversary behavior modeling based on the MITRE ATT&CK matrix.

Al-Shaer et al. propose an approach for establishing associations between the MITRE ATT&CK techniques based on hierarchical clustering and demonstrate that it could be used to forecast an attacker's behavior [6]. In [10] Kim et al. apply the MITRE ATT&CK matrix to perform attacker attribution, they developed an automated system that takes Indicators of Compromise and MITRE ATT&CK matrix in the vectorized form to attribute mobile threat actors, and show that it is possible to achieve high efficiency in defining threat actor.

The approach proposed in the paper is close to one suggested in [13], however, it performs security incident mapping to attack patterns in real-time mode and considers the step of the event correlation and construction of the security incidents and alerts.

---

[2] https://www.ibm.com/qradar/security-qradar-siem.

[3] http://www.microfocus.com/en-us/cyberres/secops/arcsight-esm.

[4] https://www.ptsecurity.com/ww-en/products/mpsiem/.

[5] https://mitre.ptsecurity.com/en-US/techniques?utm_source=pt-main-en&utm_medium=slider&utm_campaign=mitre.

3 Technique for Mapping the Security Events to the Attack Patterns

The paper proposes a technique for security incident detection using the correlation rules with their further mapping to the indicators of attacks, i.e. to the stages of the targeted cyber attacks (Fig. 1).

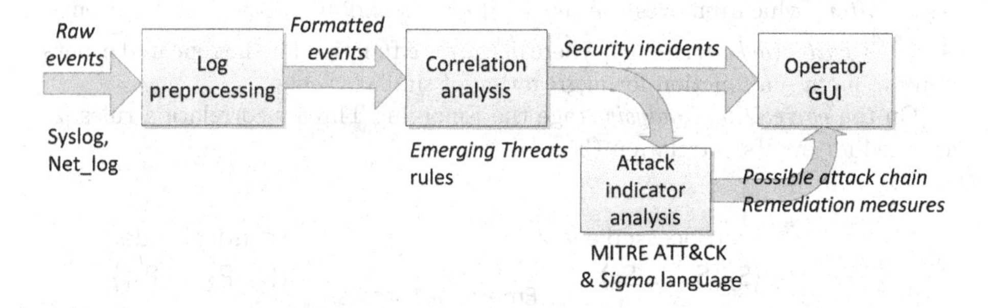

Fig. 1. The generic scheme of the proposed technique.

The proposed technique incorporates the following stages: data gathering from various sources; log preprocessing stage – this stage includes format normalization, filtering and aggregation, and data preprocessing; correlation analysis; attack indicators analysis; and results representation.

On the *data gathering stage* the raw data from the network log *net_log* and from the internal log *syslog* are fed to the preprocessing stage.

On the *preprocessing stage* the raw data A_r are normalized, preprocessed, filtered and aggregated (Fig. 2).

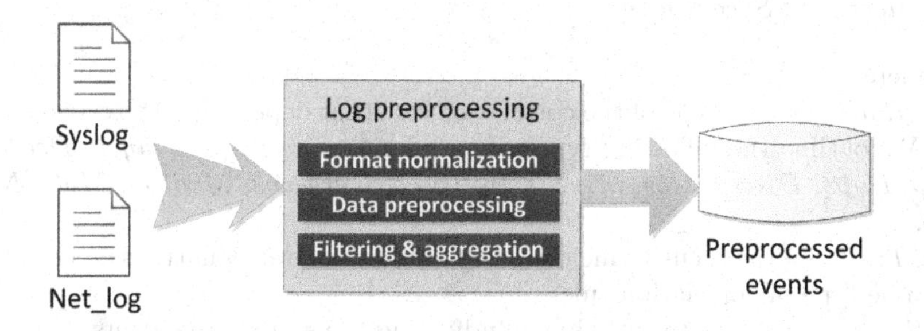

Fig. 2. The scheme of the log preprocessing process.

On this stage the following algorithm is used:

Step 1: the set of network and internal events A_r enter the normalization process ($Norm$). The events are converted to the normalized format in terms of length and syntax: $A_r \xrightarrow{read} Norm(len, syn)$, where len – a function specifying the fixed length, syn – a function specifying the normalized event syntax.

Step 2: the normalized events enter the preprocissing process ($Proc$) where the events are supplemented by the fields essential for the correlation ($time\_start$, $time\_end$, $list$): $A_r \xrightarrow{read} Proc(time\_start, time\_end, list)$, where $time\_start$ – start time of the event; $time\_end$ – end time of the event; $list$ – event source.

Step 3: the preprocessed events enter the filtering and aggregation process ($Filter$) which removes the repeated events and aggregates similar events: $A_r \xrightarrow{read} Filter(del\_atr, meta)$, where del_atr – a function for the repeated events remove; $meta$ – a function for aggregation of similar events.

On the *correlation analysis* stage the Emerging Threats correlation rules [3] are used for events correlation (Fig. 3).

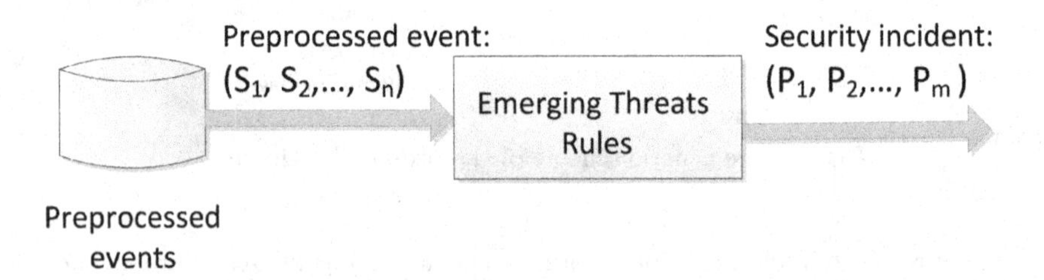

Fig. 3. The scheme of the log correlation analysis.

We understand the rule as the following mathematical object (the production model that uses IF-THEN rules to represent operation):

$$< Rule\_type > [< TriG > (S_1, S_2, ..., S_n) \xrightarrow{\text{Impact}} (P_1, P_2, ..., P_k) \xrightarrow{\text{Display}} < Alert >, < Severity >],$$

where:

$< Rule\_type >$ – type of the correlation rule that depends on the event source $list$. We outline the following types of correlation rules: $App - layer$, $Decoder$, $Dhcp$, $Dnp3$, DNS, $Files$, $http2$, $http$, $Ipsec$, $Kerberos$, $Modbus$, $Mqtt$, Nfs, Smb, Tls;

$< TriG >$ – the security incidents signatures, several signatures can exist for the same type of correlation rule;

$(S_1, S_2, ..., S_n)$ – event attributes indicating the security incidents;

$(P_1, P_2, ..., P_k)$ – detected security incidents;

$< Alert >$ – alert on the cyber security incident;

$< Severity >$ – severity of the alert (low, medium, or high).

The set of the preprocessed events A enter the correlation analysis (Fig. 3). The correlation rule is selected depending on the event source $list$. The event attributes S trigger alert $Alert$ if they correspond to one or several signatures

TriG of the security incidents P. The severity of the alert depends on the number of security incidents. It can be specified as follows:

Step 1: event enters the correlation rule depending on its source *list*: $a_i \xrightarrow{read} Rule\_type$.

Step 2: map event syntax (event attributes) S with features specified within the cyber security incident signature *TriG*: $a_i \xrightarrow{read} < Rule\_type >< TriG > (S_1, S_2, ..., S_n)$.

Step 3: if the event a_i contains at least one attribute that corresponds to the feature specified in the cyber security incident signature using the correlation rule then the incident is detected: $a_i \xrightarrow{read} < Rule\_type >< TriG > (S_1, S_2, ..., S_n) \xrightarrow{Impact} (P_1, P_2, ..., P_k)$.

Step 4: the alert *Alert* is generated and sent to the operator's GUI together with its severity: $< Rule\_type > [< TriG > (S_1, S_2, ..., S_n) \xrightarrow{\text{Impact}} (P_1, P_2, ..., P_k) \xrightarrow{\text{Display}} < Alert >, < Severity >]$.

On the *attack indicators analysis* stage the correlated events (security incidents) are mapped to the cyber attack stages according to the MITRE ATT&CK tactics and techniques (i.e. to the stages of the targeted cyber attacks) [2]. For this goal a set of TAA (IOA) rules [1] is used. These rules describe security incidents (signatures) using Sigma language [4]. The IOA allows mapping the security incidents (signatures) to the attack stages specified in the MITRE ATT&CK database [2], namely, reconnaissance, resource development, initial access, execution, persistence, privilege escalation, defense, evasion, credential access, and discovery.

This mapping can be specified as follows:

$< Sign > [< attack > (P_1, P_2, \ldots, P_n) \xrightarrow{Display} < 0, 1 >, < R, MITRE_{Obj} >]$,

where: $< Sign >$ – TAA (IOA) signatures of the security incidents;

$< attack >$ – attack techniques according to the MITRE ATT&CK. It can take the following values: $Ddos\_attack$, $malv\_attack$, $Scan\_attack$, Web_attack, Sql_attack, XSS_attack, $Shell\_attack$, Dos_attack, $Brut\_attack$, $Pass\_attack$, $Inject\_attack$;

$(P_1, P_2, \ldots, P_n)$ – cyber security incidents corresponding to the attack technique;

$< 0, 1 >$ – the result of mapping of the security incidents $(P_1, P_2, \ldots, P_n)$ to the MITRE ATT&CK techniques $< attack >$ based on the signature $< Sign >$: 0 – the set of the detected incidents $(P_1, P_2, \ldots, P_n)$ do not correspond to the $< Sign >$, 1 – the set of the detected incidents $(P_1, P_2, \ldots, P_n)$ correspond to the $< Sign >$;

R – risk for the attack;

$MITRE_{Obj}$ – attack description, its stage, possible next steps, and attack responses according to the MITRE ATT&CK.

Namely, the signatures of the detected on the correlation stage security incidents are compared with the TAA (IOA) signatures of the same incidents. If they map, then the MITRE ATT&CK targeted attack is detected, otherwise, the detected incident can't be mapped to the multi-step targeted attack. Depending on the attack stage as well as used tactics and techniques the attack responses differ. Mapping of the security incidents to the attack stages will allow prospective attack response selection in the future.

The algorithm developed for this stage can be specified as follows:

Step 1: comparison of the signatures of the detected on the correlation stage security incidents with TAA (IOA) signatures: $a_i(P_1, P_2, \ldots, P_k) \xrightarrow{compare} < Sign >< attack > (P_1, P_2, \ldots, P_n)$.

Step 2: if the signatures map then the system displays an attack risk R and $MITRE_{Obj}$: $a_i(P_1, P_2, \ldots, P_k) \xrightarrow{compare} < Sign > [< attack > (P_1, P_2, \ldots, P_n) \xrightarrow{display} < 1 >, < R, MITRE_{Obj} >]$.

Step 3: if the signatures do not map then the security incident can not be mapped to the MITRE ATT&CK stages.

4 Implementation and Experiments

The proposed technique was implemented using Python programming language and the Flask framework[6]. The general architecture of the developed tool is represented in Fig. 4.

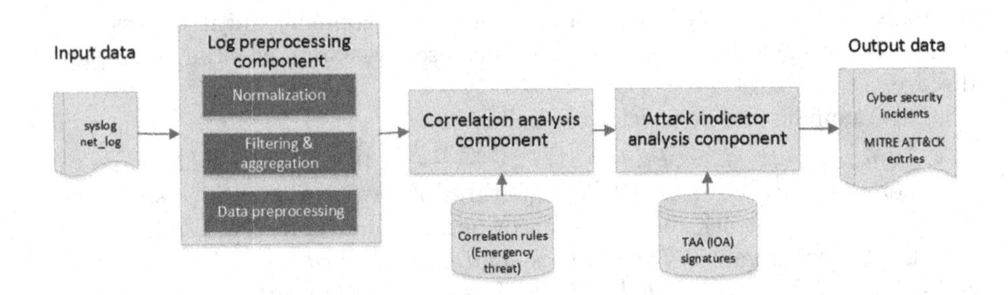

Fig. 4. Components of the developed tool.

For the experiments the testing environment was deployed (Fig. 5). The authors conducted the attacks against the user's workstation (Table 1). The tool implementing the proposed approach was installed on the Administrator's workstation as well as the tested SIEM tools.

The attacks were conducted using internal tools of the Kali Linux operation system[7] (Table 1). The results of the experiments are provided in Table 2: the target IP address, the conducted attacks, the types of events corresponding to the attacks, the TAA signatures corresponding to the attacks, the administrator IP address, and if the attack was detected and mapped.

[6] https://flask.palletsprojects.com/en/2.1.x/.
[7] https://www.kali.org/.

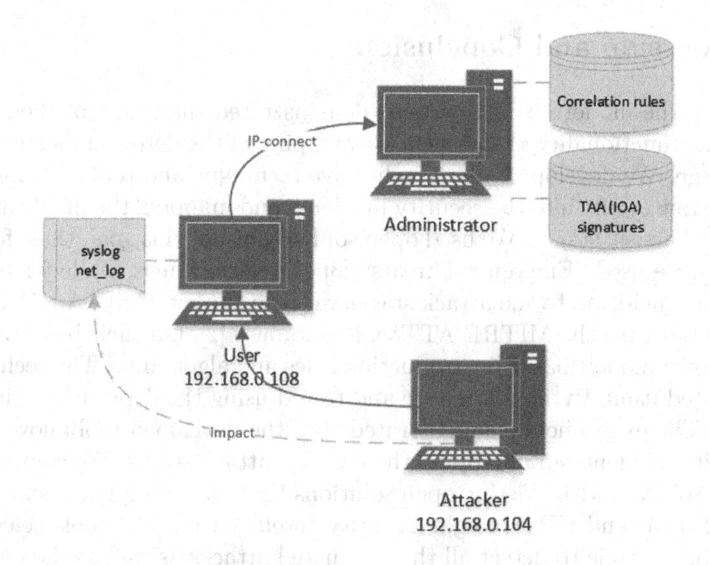

Fig. 5. Test environment.

Table 1. The conducted cyber attacks

Source IP address	Attack	Target IP address	Implemented
192.168.0.104 (attacker)	Dos_impact	192.168.0.108 (user)	✓
192.168.0.104 (attacker)	Shell_impact	192.168.0.108 (user)	✓
192.168.0.104 (attacker)	Scan_impact	192.168.0.108 (user)	✓
192.168.0.104 (attacker)	Inject_impact	192.168.0.108 (user)	✓
192.168.0.104 (attacker)	Brut_impact	192.168.0.108 (user)	✓

Table 2. The detected and mapped attacks

User IP address	Attack	Event type	TAA signature	Administrator IP address	Detected
192.168.0.108 (user)	Dos_impact	C	Dos_attack	127.0.0.1	✓
192.168.0.108 (user)	Shell_impact	L	Shell_attack	127.0.0.1	✓
192.168.0.108 (user)	Scan_impact	C	Scan_attack	127.0.0.1	✓
192.168.0.108 (user)	Inject_impact	L	Inject_attack	127.0.0.1	✓
192.168.0.108 (user)	Brut_impact	C & L	Brut_attack	127.0.0.1	✓

5 Discussion and Conclusion

Analysis of the modern SIEM systems demonstrated that most of them do not provide the functionality of the accurate mapping of the detected incident to the attack stage. We developed a comprehensive technique and tool allowing correlating the raw events into the security incidents and mapping the incidents to the attacks and attack stages. We used open source tools for this goal, thus, for event correlation we used "Emerging Threats" [3] correlation rules, while for mapping the security incidents to the attack stages we used the set of "TAA (IOA)" rules [1] integrated with the MITRE ATT&CK database [2]. On their basis we developed our own models based on production rules and algorithms. The technique is implemented using Python language and tested using the deployed testing environment. The experiments demonstrated that the developed tool allows detecting security incidents and mapping them to the attack stages. We compared the proposed solution with existing open solutions (Splunk Enterprise Security, IBM QRadar SIEM, and HP ArcSight Security Intelligence). The tools selected for comparison are able to detect all the conducted attacks as well as the developed tool (Table 3). But unlike existing tools the developed tool is also able to detect techniques and tactics corresponding to the detected incident according to the MITRE ATT&CK (Table 3). In future work the authors plan to use more complex correlation rules for the detection of cyber security incidents to cover more cyber attack scenarios and extend the experiments to map other types of cyber security incidents to the MITRE ATT&CK tactics and techniques.

Table 3. Comparison of the developed tool and existing SIEMs

Attack	Dos_attack	Scan_attack	Shell_attack
Attack detected			
Developed tool	✓	✓	✓
Other tools	✓	✓	✓
Attack mapped			
Developed tool	T1499.001[1] T1499.002[2] T1499.003[3] T1499.004[4]	T1595.001[5] T1595.002[6]	T1505.001[7] T1505.002[8] T1505.003[9]
Other tools	✗	✗	✗

[1] https://attack.mitre.org/techniques/T1499/001/
[2] https://attack.mitre.org/techniques/T1499/002/
[3] https://attack.mitre.org/techniques/T1499/003/
[4] https://attack.mitre.org/techniques/T1499/004/
[5] https://attack.mitre.org/techniques/T1595/001/
[6] https://attack.mitre.org/techniques/T1595/002/
[7] https://attack.mitre.org/techniques/T1505/001/
[8] https://attack.mitre.org/techniques/T1505/002/
[9] https://attack.mitre.org/techniques/T1505/003/

The proposed technique adds an additional stage to the SIEM system's operation, namely, detection of the attack stage. While it requires additional time and resources, this stage is reasonable as soon as it allows automation of the cyber incident analysis and as the result saving the resources in future.

Detection of the attack stage and specific tactic or technique of the attack is essential for the targeted multi step attacks prevention. Detection of the targeted cyber attack at early stage and correct forecasting of the attack goal allow avoiding the attack success and impact from its successful implementation. In the future work we plan to develop attack forecasting technique and technique for prospective countermeasures selection.

Funding Information. This research is being supported by the grant of RSF #21-71-20078 in SPC RAS.

References

1. Kaspersky Anti Targeted Attack Platform: Indicators of compromise (IOC) and attack (IOA) for threat hunting. https://support.kaspersky.com/KATA/3.7.1/en-US/194907.htm
2. MITRE: MITRE ATT&CK Knowledge base. https://attack.mitre.org/
3. Proofpoint Inc.: Proofpoint Emerging Threats Rules. https://rules.emergingthreats.net/open/suricata-5.0/rules/
4. Sigma: SigmaHQ. https://github.com/SigmaHQ/sigma/tree/master/rules
5. Ajmal, A.B., Shah, M.A., Maple, C., Asghar, M.N., Islam, S.U.: Offensive security: towards proactive threat hunting via adversary emulation. IEEE Access **9**, 126023–126033 (2021). https://doi.org/10.1109/ACCESS.2021.3104260
6. Al-Shaer, R., Spring, J.M., Christou, E.: Learning the associations of MITRE ATT&CK adversarial techniques. In: 2020 IEEE Conference on Communications and Network Security (CNS), pp. 1–9 (2020). https://doi.org/10.1109/CNS48642.2020.9162207
7. Croxton, F., Cowden, D.: Applied General Statistics (1959)
8. Doynikova, E., Novikova, E., Gaifulina, D., Kotenko, I.: Towards attacker attribution for risk analysis. In: Garcia-Alfaro, J., Leneutre, J., Cuppens, N., Yaich, R. (eds.) CRiSIS 2020. LNCS, vol. 12528, pp. 347–353. Springer, Cham (2021). https://doi.org/10.1007/978-3-030-68887-5_22
9. Elitzur, A., Puzis, R., Zilberman, P.: Attack hypothesis generation. In: 2019 European Intelligence and Security Informatics Conference (EISIC), pp. 40–47 (2019). https://doi.org/10.1109/EISIC49498.2019.9108886
10. Kim, K., Shin, Y., Lee, J., Lee, K.: Automatically attributing mobile threat actors by vectorized att&ck matrix and paired indicator. Sensors **21**(19) (2021). https://doi.org/10.3390/s21196522
11. Kotenko, I., Fedorchenko, A., Doynikova, E.: data analytics for security management of complex heterogeneous systems: event correlation and security assessment tasks. In: Shandilya, S.K., Wagner, N., Nagar, A.K. (eds.) Advances in Cyber Security Analytics and Decision Systems. EICC, pp. 79–116. Springer, Cham (2020). https://doi.org/10.1007/978-3-030-19353-9_5

12. Kotenko, I., Gaifulina, D., Zelichenok, I.: Systematic literature review of security event correlation methods. IEEE Access **10**, 43387–43420 (2022). https://doi.org/10.1109/ACCESS.2022.3168976
13. Nisioti, A., Loukas, G., Laszka, A., Panaousis, E.: Data-driven decision support for optimizing cyber forensic investigations. IEEE Trans. Inform. Foren. Secur. **16**, 2397–2412 (2021). https://doi.org/10.1109/TIFS.2021.3054966

Author Index

Printed in the United States
by Baker & Taylor Publisher Services